Ground Anchors

Jean-Jacques Henri Rincent

Ground Anchors

Tension Force—Vibratory Analysis

 Springer

Jean-Jacques Henri Rincent
Rincent Gestion
Évry-Courcouronnes, Essonne, France

ISBN 978-981-97-4413-8   ISBN 978-981-97-4414-5 (eBook)
https://doi.org/10.1007/978-981-97-4414-5

This work was supported by Rincent Gestion.

This Springer imprint is published by the registered company Springer Nature Singapore Pte Ltd.
The registered company address is: 152 Beach Road, #21-01/04 Gateway East, Singapore 189721, Singapore

If disposing of this product, please recycle the paper.

Preface

Tension in the tie rods decreases with time and also with the type of load. Managers of railroad networks transporting ore, or road concessionaires, are wondering about the stability of retaining walls provided by tie rods that have been in place for several decades. The method described in this book has already been in operation for 20 years, and has taken on a new dimension with the Brazilian market. The number of tie rods is very large, which meant that from 2019 to 2024, several thousand non-destructive tests were carried out to confirm and develop this diagnosis.

Initially, the internal tension force of the tie-rod is calculated from the dynamic stiffness obtained from vibration analysis tests. The method also determines the total length of the tie-rod and its free length. Finally, the tie's mobility response is used to calculate its diameter. All these elements are used to resize the tie rods, and then to calculate the stability of the retaining walls.

The next point concerns the possibility of re-tensioning the tie rods. When a non-destructive test is carried out on tie rods, 8 tests are actually performed. Over the last 5 years, 4,000 tie rods have been tested, representing 32,000 tests. These results, analyzed individually, represent a wealth of experience.

This procedure enables us to analyze the stability of the dynamic stiffness. The evolution of dynamic stiffness with each hammer strike reveals fragile tie rods.

The cost of re-tensioning is 10 to 30 times lower than that of a new tie bar. Some re-tensioning operations have been carried out, but it's the mechanical work involved that's the most complicated, as the heads of the tie rods don't always allow for easy re-tensioning. Access to testing and re-tensioning operations must be a priority for those involved in this particular field.

During these tests, we noted the negative effect of cyclic loads induced by convoy or heavy vehicle traffic. This raises the question of how to take this type of stress into account in the design phase, including during reinforcement.

Évry-Courcouronnes, France   Jean-Jacques Henri Rincent

Introduction

The book presents a non-destructive vibration analysis method for calculating the internal tension force of the tie rods tested.

This method was formalized in 2003, and after 2 years of investigation was patented, our patent which expires in 2025.

The aim of this book is therefore to present this method, based on numerous concrete examples.

The theory and principles of instrumentation are explained. Analysis of the vibratory responses enables us to calculate the free and total lengths of the tie rod, as well as the diameter of the tie rod, i.e., the reinforcement surrounded by its grout.

The strength of this document is that it shows the relationship between dynamic stiffness and the tension force of the tie-rod, and the method for calibrating with static tensile tests.

Static tensile tests are difficult to perform at height, due to the weight of the equipment. These static tests present a real risk of failure, particularly on older tie rods.

The examples presented come mainly from tests carried out in Brazil, the old structures, which existed for several decades, show load losses over these periods ranging from 1% to 5.5% per year.

On one of the walls tested, a limited number of tie rods were re-tensioned.

Retaining wall managers are increasingly using this method to carry out maintenance diagnostics. This method enables a large sample to be taken, which is representative of the existing situation.

Contents

About the Author

Dr. Jean-Jacques Henri Rincent Doctor of Civil Engineering (Institute National des Sciences Appliquées of Toulouse INSA) Chairman and CEO of Rincent BTP Services.

The main lines of development are:

- Determination of the anchors tension force with vibratory analysis.
- Dynamic Behavior of Pavement under Impact Loading.
- Development of a procedure for analyzing the bitumen content of asphalt mixes with the Ground Penetration Radar.

Former Director of the Geomechanics Department of the CEBTP (Experimental Center for Building and Public Works) for 10 years, Associate Assistant Professor Polytech Lille Civil Engineering section for 2 years and at the start teacher at the Civil Engineering Department of the National Engineering School of LIBREVILLE Gabon, School linked by contract to INSA of Toulouse.

Chapter 1
Waves-Vibratory Analysis

Radio waves are electromagnetic waves of the same nature as light, which result from the disturbance of electric and magnetic fields.

In addition to visible light, there are radar waves, microwaves, infrared, ultraviolet, X and gamma rays, and radio waves.

Our area of intervention is sound. Of sound waves need material support to spread, electromagnetic waves, spread better in a vacuum, and much faster: sound only goes at 300 m/s, while electromagnetic waves go at about 300,000 km/s. You see lightning before you hear thunder.

1.1 Definitions—Period—Frequency

In addition to visible light, there are radar waves, microwaves, infrared, ultraviolet, X-rays, gamma rays and radio waves. Our operating range is generally below 2000 Hz.

A wave is the modification of the physical state of a physical environment. In our case, it's a reinforcement, usually steel, contained in a cylinder of cement grout. The wave propagates as a result of local action, with a velocity determined by the characteristics of the material through which it passes.

A periodic function is a function which, when applied to a variable, takes on the same value as a certain fixed quantity called period is added to that variable (Fig. 1.1).

The period T is the time between the passage of two maxima.

Frequency f is a measure of the number of times a periodic phenomenon repeats per unit time $f = 1/T$ (Fig. 1.2).

The frequency is the number of periods per unit time, which is the inverse of the period: $f = 1/T$ where f is the frequency in Hertz (Fig. 1.3).

© The Author(s) 2024
J.-J. H. Rincent, *Ground Anchors*, https://doi.org/10.1007/978-981-97-4414-5_1

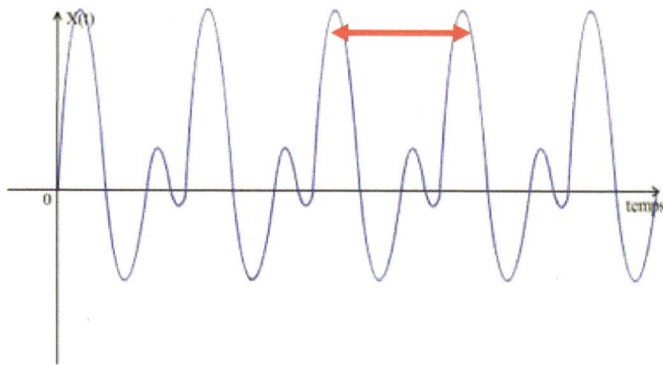

Fig. 1.1 Wave / period "T" example

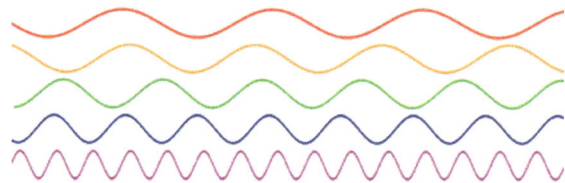

Fig. 1.2 Sinusoidal waves of different frequencies. — the lower one has the high frequency. — the upper one has the lower one.

Fig. 1.3 Types of frequency

1.2 Principle of the Method

The principle of the method is described in:

– European Patent Office PCT/FR2005/001597 (Appendix A)
– WIPO Patent No. WO 2006/010830 A1, 2006 (Appendix B)

  The method consists of:

– Delivering a mechanical shock to the head of the tested element by means of a hammer equipped with a force sensor.
– Measure the particle velocity at the head of the test item.
– Do the acquisition in time mode.
– And analyze the signal in frequency mode.

Fig. 1.4 Diagram of the compression wave

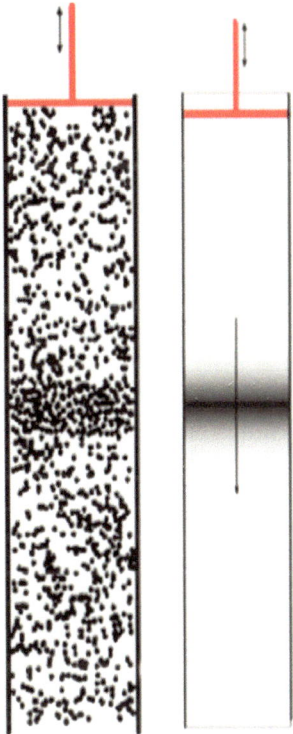

The figure below visualizes the compression wave emitted by the impact of the hammer and which propagates inside the tested element to go up at its extremity (Fig. 1.4).

1.3 Mechanical Impedance

Impedance in a circuit fed by an alternating current is equivalent to resistance.

It is the factor that absorbs the energy used.

Whatever it is:

– Electrical
– Mechanical
– Acoustics

Mechanical impedance is a measure of the resistance to motion of a structure subjected to a given periodic force. It refers to the velocity forces acting on a mechanical system. The mechanical impedance of a point relative to a structure is the relationship between the force applied at a point and the resulting velocity at that point. Mechanical impedance is the inverse of mechanical admittance or mobility.

Fig. 1.5 V/F curve as a
function of frequency

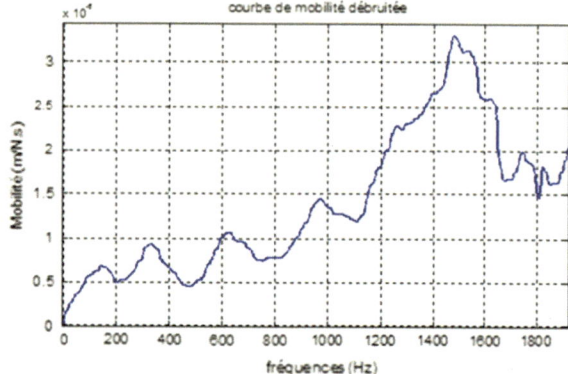

Fig. 1.6 Vibratory
responses

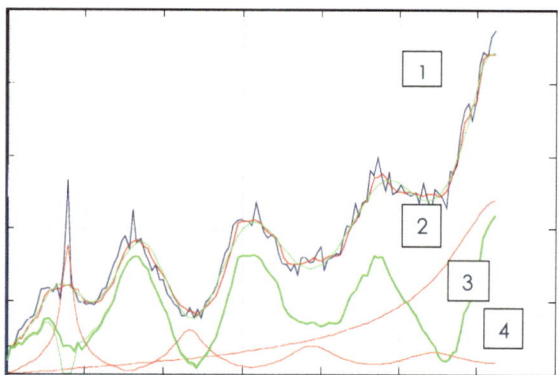

Example of a curve resulting from a mechanical impedance test:

F in Newton V in m/s V/F in m/sN, frequency in Hz (Figs. 1.5 and 1.6).

The blue test curve (1) is the geometric sum of the three green (3) and two red curves (2 and 4). These curves correspond to different vibration responses of the tested element.

1.4 Ties Rods

A tie rod has

- A high strength steel bar or strand wire
- A free part and a sealed part
- A cement grout to protect the tie rod and ensure the bond with the soil (Figs. 1.7 and 1.8).

Fig. 1.7 Cleaning the head and fixing the geophone 3D. *Source* Rincent BTP—Recife

Fig. 1.8 Example: free
length and sealed length

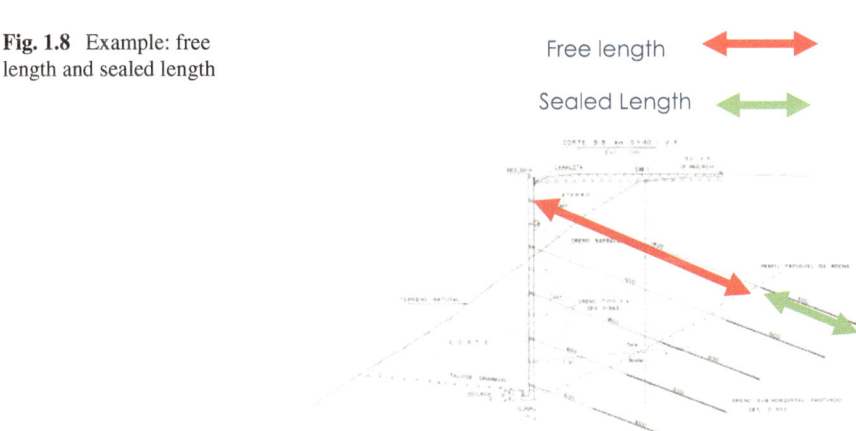

Equipment—Force and Velocity (Fig. 1.9).

The force sensor is piezoelectric.

Piezoelectricity is the property of certain materials to polarize electrically under the action of a mechanical stress and, conversely, to deform when an electric field is applied.

Equipment measures millivolts as a function of time (Fig. 1.10).

Geophone velocity measurement.

The principle of operation of the geophone is that the movement of the support on which it is fixed causes the magnet in the solenoid to move, which creates an electric

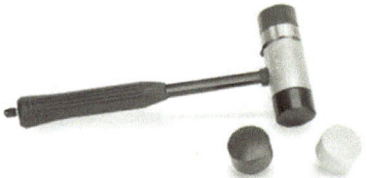

Fig. 1.9 Hammer with different ends. *Source* Rincent BTP—France

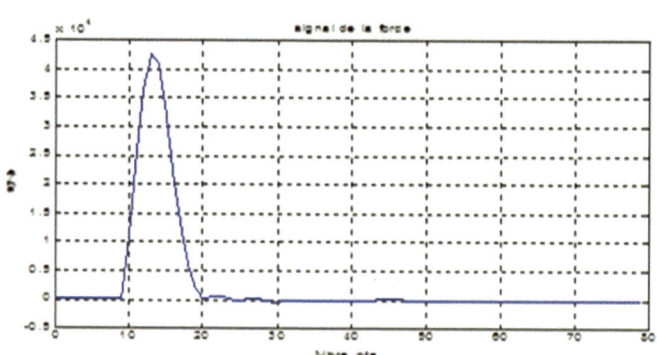

Fig. 1.10 Force versus time

current linked to the movement. Without movement there is no current measurement (Fig. 1.11).

For velocity measurement equipment is the geophone 3D or accelerometer (Figs. 1.12, 1.13 and 1.14).

Changing from temporal mode of acquisition to frequency mode for analysis of results. The Fourier transform is an operation that allows non-periodic signals to be represented in frequency.

The calculation method that allows you to pass in a reversible way from a function to the corresponding trigonometric series is the Fourier transform. This very fruitful method has become a must-have in signal theory, with important applications for the processing and compression of sound and digital images, JPEG image compression, or the 3G and 4G telephony standards are derived directly.

Fig. 1.11 Geophone, working principle. *Source* Rincent BTP—France

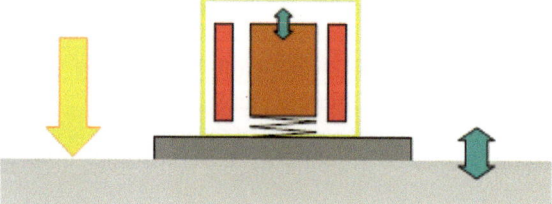

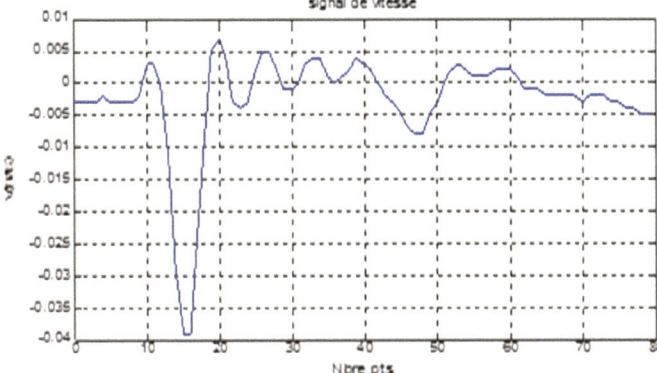

Fig. 1.12 Velocity as a function of time

Fig. 1.13 Realization of the impact. *Source* Rincent BTP—Recife

Fourier Transform.

Any curve can be decomposed into a sum of sinusoidal curves of different frequencies with a different weighting coefficient. This operation allows you to switch from time mode to frequency mode (Figs. 1.15 and 1.16).

Fourier Transform for Velocity (Fig. 1.17).

The next step is to establish V/F as a function of frequency (Fig. 1.18).

After 2000 Hz the force function of frequency is close to zero. V/F curve shows two parts one before 2000 Hz and one after 2000 Hz (Fig. 1.19).

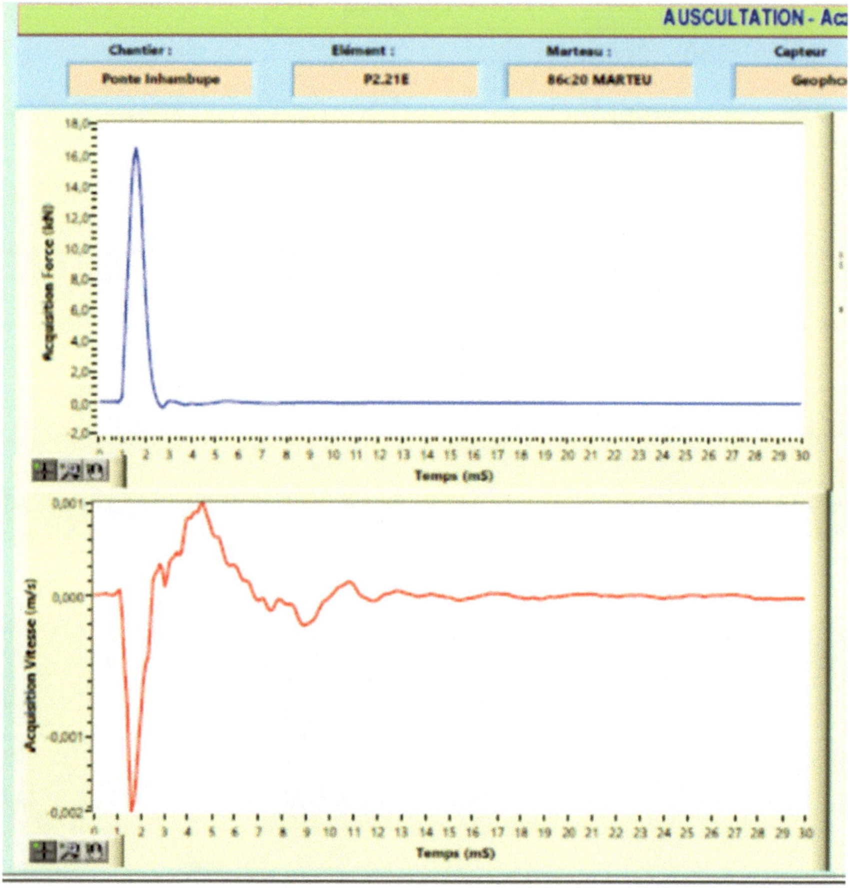

AUSCULTATION - Ac
Chantier :
Ponte Inhambupe
Elément :
P2.21E
Marteau :
86c20 MARTEU
Capteur
Geopho
Acquisition Force (kN)
Temps (mS)
Acquisition Vitesse (m/s)
Temps (mS)

Fig. 1.14 Force and velocity as a function of time test results

Fig. 1.15 Fourier transform

Σ
temps
t
ω
2 ω
3 ω
pulsation
(fq)

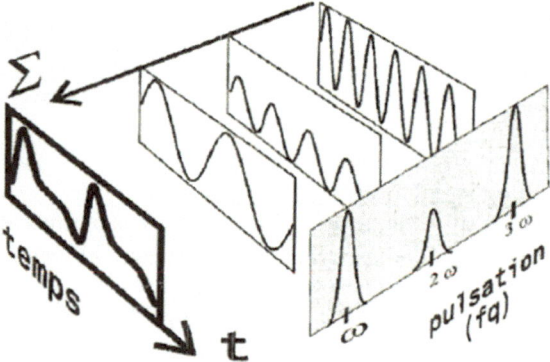

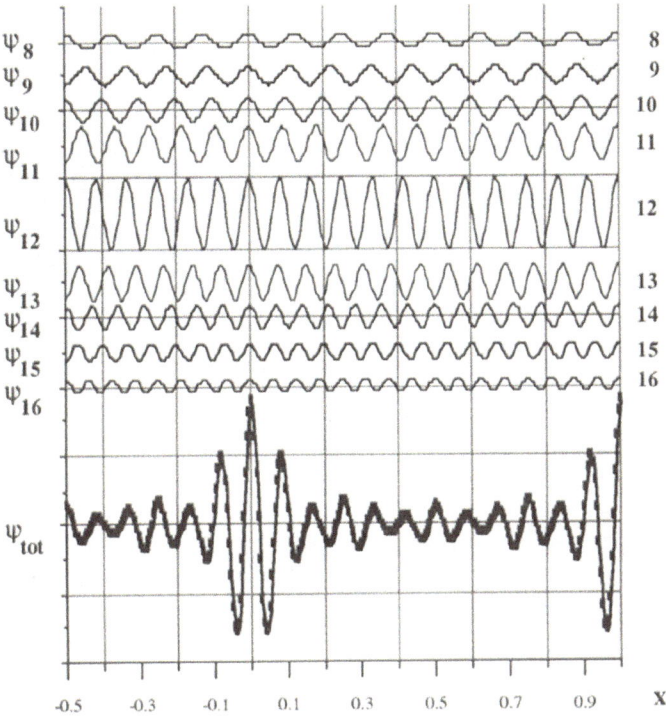

Fig. 1.16   Non-periodic curve and sum of periodic functions

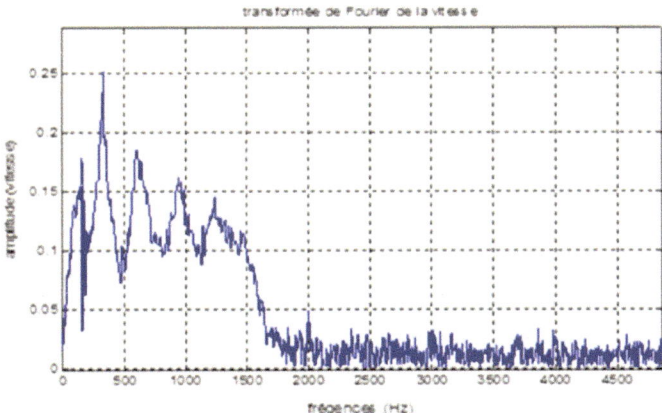

Fig. 1.17   Fourier transform for velocity

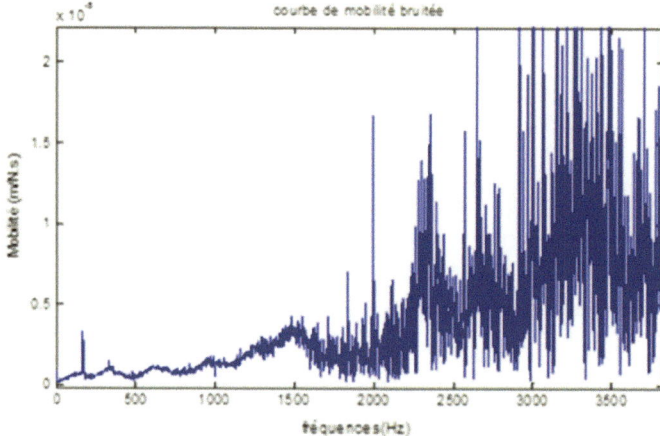

Fig. 11.18 V/F curve as a function of frequency

Fig. 1.19 V/F curve function of frequency filtered curve

Time x Frequency

A short phenomenon in time covers wide frequency ranges. In the case of thunderstorms and lightning, the electrical discharge is a short phenomenon that emits over a wide frequency range, causing disruption to radio and TV signals.

For an earthquake the phenomenon is long in a limited frequency range of 2 to 4 Hz.

A hammer with a metal end will produce in a frequency range of 0 to 5000 Hz, and with a "matte" impact elastomer end from 0 to 2000 Hz.

For us, the latter option we used.

Evaluating the V/F curve as a function of frequency

Fig. 1.20 Calculating tie
rod length

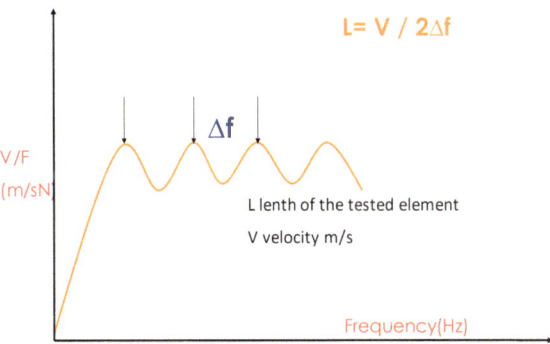

The first step is to calculate the length of the tested element (Fig. 1.20).

The value of the plane wave propagation velocity in concrete for calculation is 4000 m/s. This theoretical value is given by the testing standard.

There can be two vibration responses that will correspond to the total part of the tie and the part of soils with low mechanical characteristics found in the first meters of the tie (Fig. 1.21).

Example of the tie rod: Free part and total length of the tie rod (Figs. 1.22 and 1.23).

Overall length:

– 23,1m with a velocity assumption of 3500m/s
– 26,0m with a velocity assumption of 4000m/s (Fig. 1.24)

Length free part:

– 8.4m with a velocity assumption of 3500m/s
– 9.5m with a velocity assumption of 4000m/s

Mobility.

The mobility read directly on the y-axis allows you to calculate the diameter of the tested element (Fig. 1.25).

Fig. 1.21 Two vibration
regimes

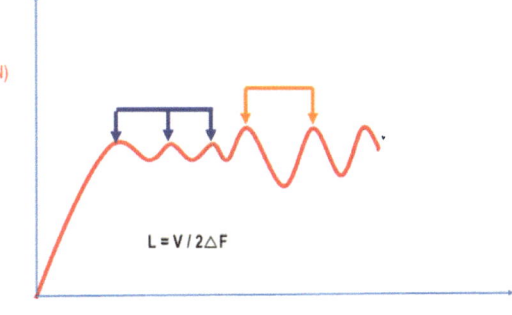

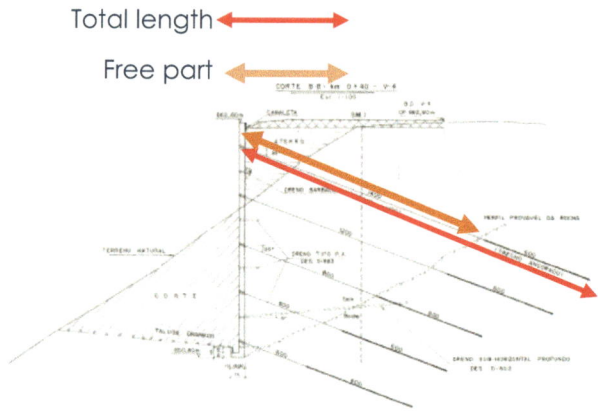

Fig. 1.22 Tie rod, free part and total tie rod

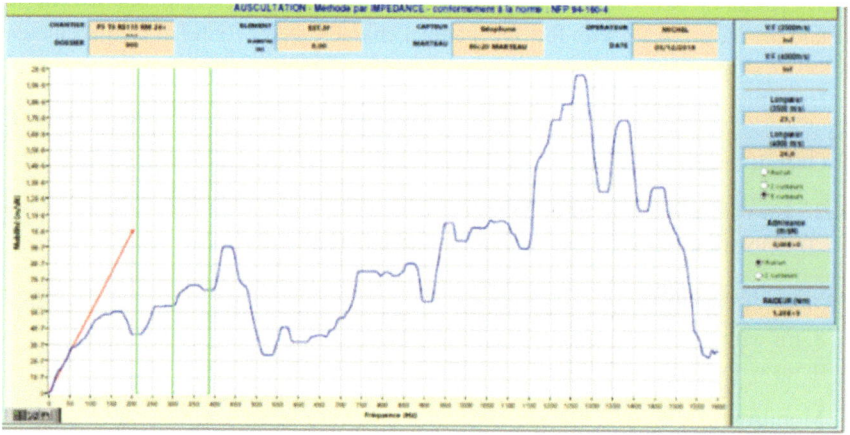

Fig. 1.23 Analyzing test results

V/F = 1 / ρb V b A

Concrete volume mass ρb in kg/m^3

Circular cross section (m^2)

Dynamic stiffness (Fig. 1.26).

Dynamic stiffness equals 2πb/a and is a complex number.

A complex number is of the type z = x + iy, X is real, Y is imaginary (Fig. 1.27).

Simultaneous static testing and non-destructive testing allow the following figure to be drawn (Figs. 1.28 and 1.29).

Real and imaginary part of stiffness.

A thesis "Measurements and Modelling of Resilient Rubber Rail-pads" by Klaus Knothe, MinyiYu, Mrs. HeikeIlias shows that stiffness part real and part imaginary is constant before 80 Hz (Fig. 1.30).

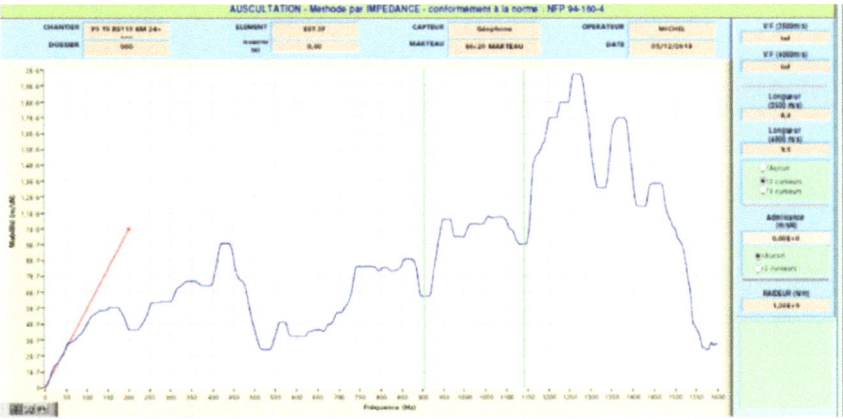

Fig. 1.24 Same test, second vibration regime

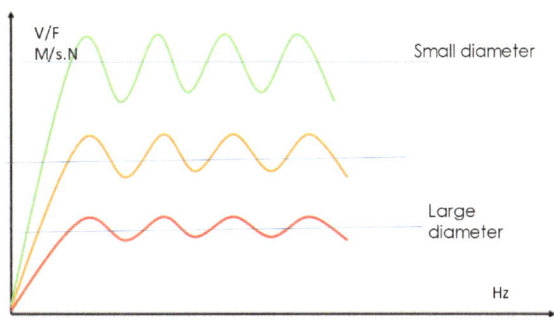

Fig. 1.25 Mobility as a function of diameter of the tested element

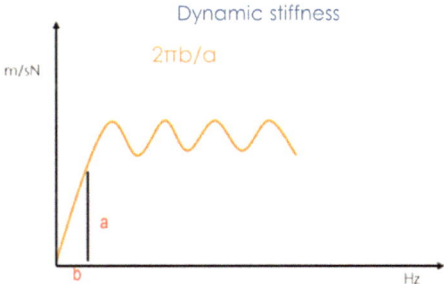

Fig. 1.26 Dynamic stiffness calculation

Static stiffness.

Static stiffness of an element under a load is the slope of the calculated curve under this load (Fig. 1.31).

Philippe Guillermain's thesis shows that there is a direct link between dynamic stiffness and static stiffness, dynamic stiffness is higher than static stiffness, there is a constant relationship between the two stiffnesses for an identical situation.

Fig. 1.27 Simultaneous
static and dynamic testing.
Source Rincent
BTP—France

Fig. 1.28 Dynamic strength
and stiffness

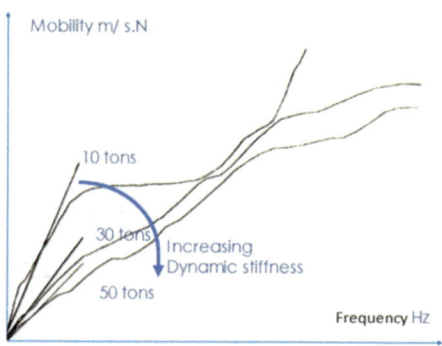

Fig. 1.29 Example of a
stiffness versus force curve
(Tons)

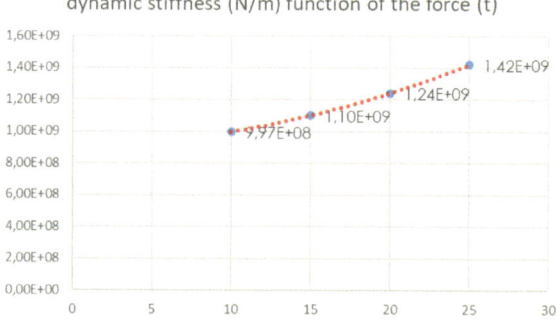

Stiffness has the same units as modulus.

The Rd/Rs calculations were done from tests on tie rods, the relationship between dynamic and static stiffness and 31 by following example (Fig. 1.32).

This rule can be checked for other elements, e.g., micro piles in tension (Fig. 1.33).

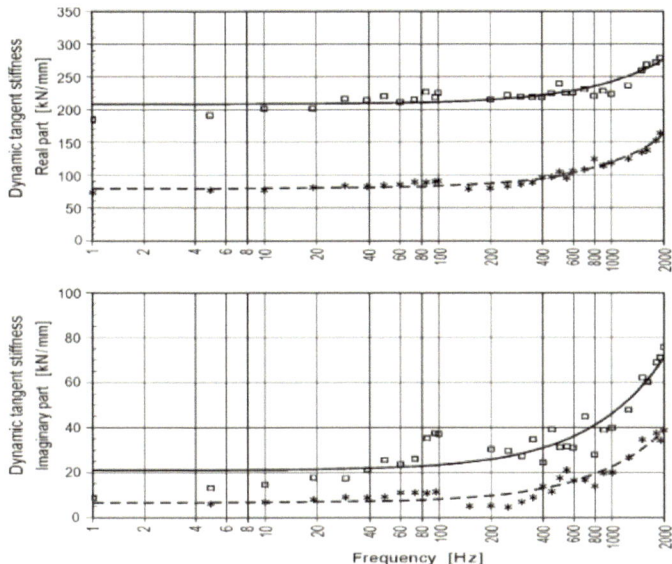

Fig. 1.30 Dynamic stiffness real part and imaginary part

Fig. 1.31 Static stiffness
under 3000 kN load

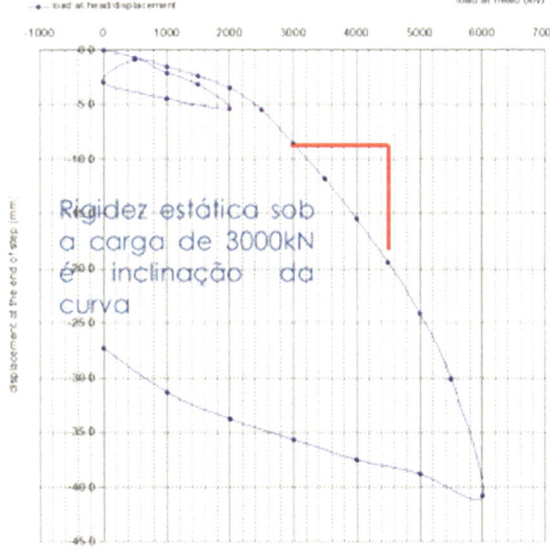

Rigidez estática sob
a carga de 3000kN
é inclinação da
curva

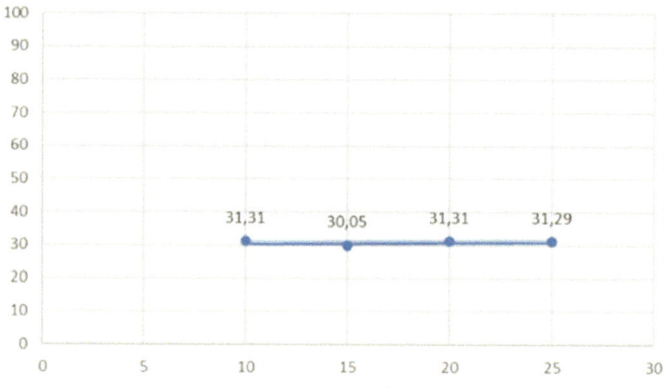

Fig. 1.32 Rd/Rs for different loads (tons)

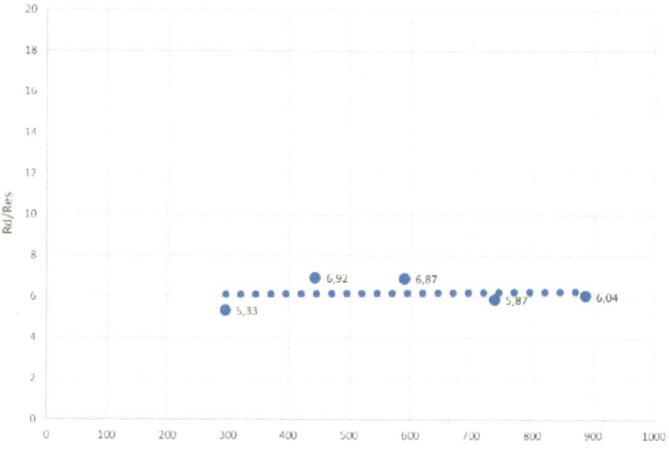

Fig. 1.33 Rd/Rs for different loads (kN)

Chapter 2
Phases of Work

2.1 Preliminary Observations

The examples presented in this book are based on projects carried out in Brazil. The heads of the tie rods are protected by concrete blocks which cover the distribution plate in contact with the wall. The geophone is generally glued to this metal plate, and hammer blows are made on the head of the tie rod itself. It is therefore necessary to destroy the concrete protection in order to carry out the tests.

The measured stiffness is the sum of the tie rod inertia of the wall and the tensile force of the tie.

The quality of the dynamic tests is linked to the quality of the mounting of the sensor that measures the velocity. Several solutions are possible, for example, gluing a metal plate that allows the geophone to be fixed by bolting, the vibration response is that of the tie rod, not the sensor.

The quality of the bond must withstand 8 consecutive tests on the head of the tie rod, over which the dynamic stiffness value is calculated, the maximum and minimum values are eliminated, the average is calculated over the remaining 6.

The geophone fastening is the most important point in this work, a poorly performed test is an unusable result.

2.2 Accessibilities

See Fig. 2.1.

© The Author(s) 2024
J.-J. H. Rincent, *Ground Anchors*, https://doi.org/10.1007/978-981-97-4414-5_2

Fig. 2.1 Access to the wall
and clearing. *Source* Rincent
BTP—Recife

2.3 Preparation of Tie Rods for Testing

In addition to cleaning the heads of the tie rods, the essential operation is the
numbering of the tie rods or their identification on an existing document. It is also
the opportunity to photograph everything that will help in the interpretation of the
tests (Figs. 2.2 and 2.3).

2.4 Characteristics of the Tie Rods

The first characteristic measured is the diameter of the strands, the bars and the
dimensions of the load distribution plate (Fig. 2.4).

Note the slope of the base plate and add descriptions such as water ingress and
corrosion. Photographs are essential (Figs. 2.5, 2.6 and 2.7).

2.5 Non-destructive Testing

The preparation of the tie rod head and the quality of the 3D geophone mounting are
essential operations to obtain quality results (Fig. 2.8).

Fig. 2.2 Numbering the
tie-rods. *Source* Rincent
BTP—Recife

Fig. 2.3 Destruction of the
concrete heads. *Source*
Rincent BTP—Recife

Fig. 2.4 Steel diameters. *Source* Rincent BTP—Recife

Fig. 2.5 Measurement of
the inclination of the support
plates. *Source* Rincent
BTP—Recife

Fig. 2.6 Water entering.
Source Rincent
BTP—Recife

Fig. 2.7 The embankment closes a talweg. *Source* Rincent BTP—Recife

Fig. 2.8 Securing the geophone support. *Source* Rincent BTP—Recife

The next step is to clean the metal surfaces to which the geophone will be attached. There are a number of techniques for fixing the geophone to ensure the quality of the signals recorded. By way of example, the figures below show the response of a velocity sensor correctly attached and directly attached to the same element where surface oxidation has not been removed (Figs. 2.9 and 2.10).

The amplitude of the signal is halved, and the response includes numerous parasitic vibrations.

On one dam, the disturbance was caused by the vibrations generated by the flow of water, particularly during floods, as can be seen from the curves obtained near the spillway (Figs. 2.11 and 2.12).

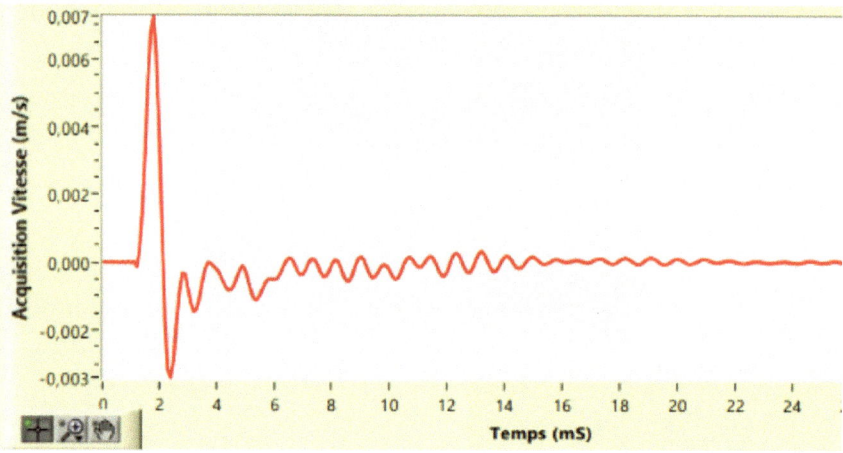

Fig. 2.9 Geophone fixed correctly

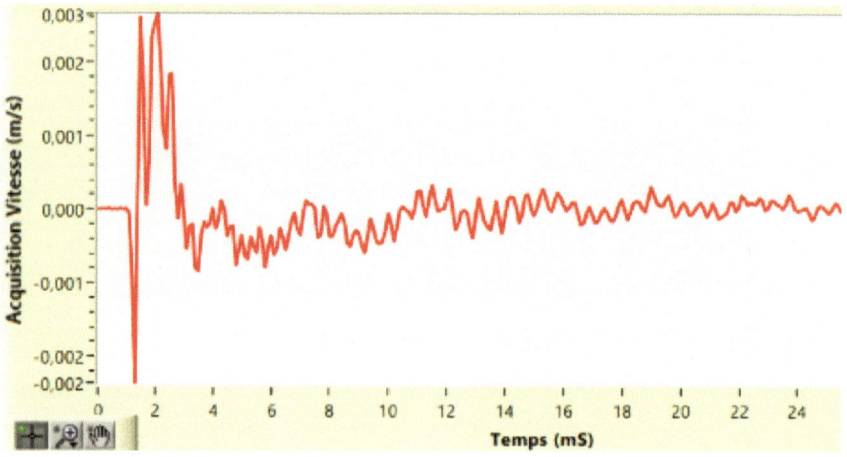

Fig. 2.10 Not fixed correctly

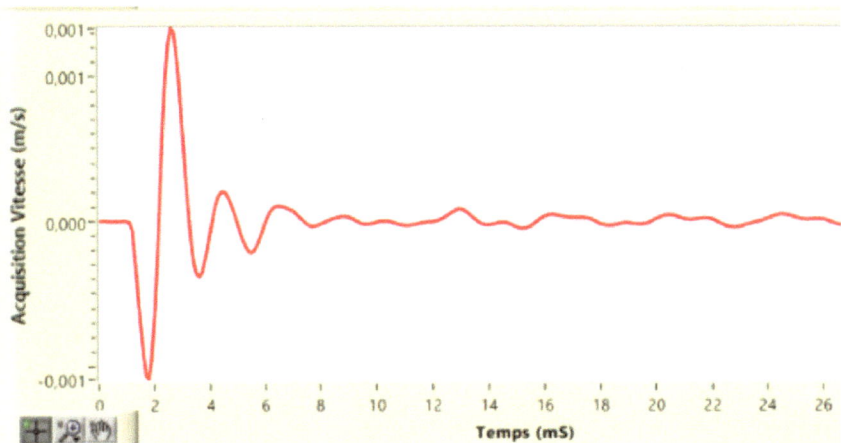

Fig. 2.11 Acquisition in calm period

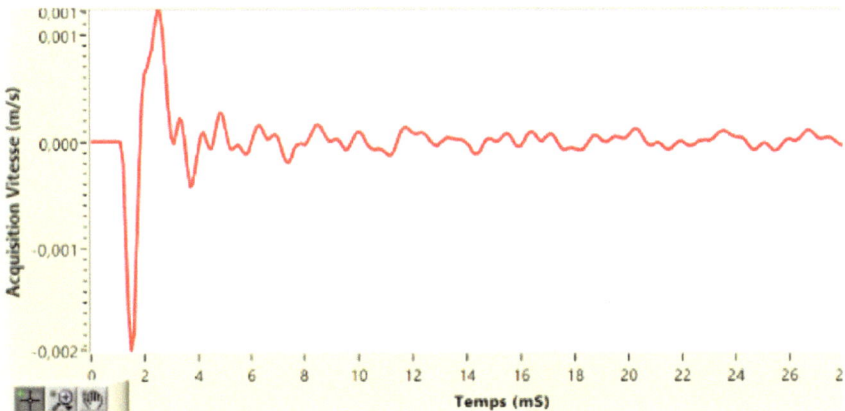

Fig. 2.12 During flood periods

This particular aspect of the geophone's mounting is essential for obtaining usable dynamic stiffness values.

Two types of acquisition were carried out on the same dam. The first procedure consists in placing the geophone on the head of the tie-rod, with the strike performed on the head of the tie-rod. The results were as follows (Fig. 2.13).

This is the response for the tie-rod with grout cement alone, and the average stiffness of 1.83E8N/m corresponds to a weight of around 2 tons, using the formula established from tests on this site. The second configuration consists of attaching the geophone to the metal distribution plate. The stiffness results are (Fig. 2.14).

The average dynamic stiffness is 4.19 E9 N/m, which allows us to calculate the internal force at 44.5 tons.

1,42E+08
1,96E+08
1,23E+08
1,64E+08
2,44E+08
2,28E+08
1,92E+08
1,75E+08

Fig. 2.13 First configuration. *Source* Rincent BTP Recife

5,08E+09
6,69E+09
5,34E+07
2,76E+09
4,71E+09
4,35E+09
3,42E+09
4,81E+09

Fig. 2.14 Second configuration. *Source* Rincent BTP Recife

Often the performance of non-destructive testing requires persons certified to carry out work at height (Fig. 2.15).

The performance of static and dynamic tests simultaneously allows you to calibrate the two methods and define the internal tensile force of the tie. The number of static tests is limited, but improves the reliability 2.14 the results. These simultaneous tests require the design and manufacture of a specific test device (Fig. 2.16).

Resistivity measurement is also a non-destructive test necessary to identify the corrosion phenomenon when it exists (Fig. 2.17).

Fig. 2.15 Non-destructive testing. *Source* Rincent BTP—Recife

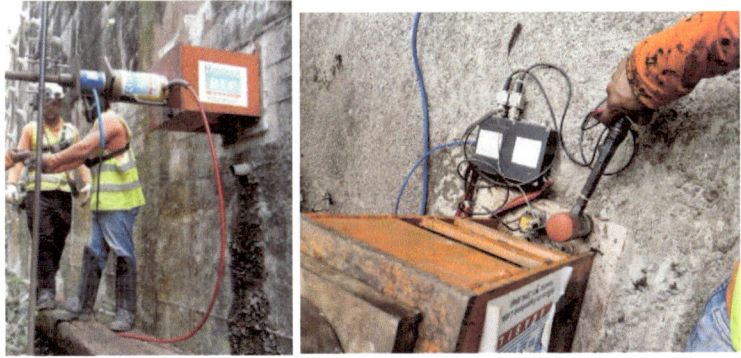

Fig. 2.16 Simultaneous static and dynamic tests. *Source* Rincent BTP—Recife

Fig. 2.17 Resistivity measurement. *Source* Rincent BTP—Recife

Chapter 3
Steels

When tie rods are 30 to 40 years old, it is necessary to conduct literature research to identify the nature of the steels. It is necessary to evaluate the initial tensile strength.

This is an important point in the identification of the tie rod.

You need to do research on the following points (Fig. 3.1):

- what year the tie rods were made
- what type of steel
- what diameter
- take measurements with a caliper
- -take samples
- and perform direct tensile tests to determine
- tensile strength
- -and Young's modulus.

Mechanical testing on steel samples taken from the test site is necessary, especially if the steels are old (Figs. 3.2 and 3.3).

The older the steels, the more important the literature search is. It is desirable to identify the origin of the steels, their place of manufacture, and their mechanical characteristics (Fig. 3.4).

Research should be directed to the calculation method used at the time of the test. Then an estimate of the value of the tie loss will allow the static tensile test to be understood with more confidence, it is necessary to design a steel structure that allows the static test to be performed simultaneously with the non-destructive test. And it is also important to know calculation methods and the calculation methods used at the time (Fig. 3.5).

An estimate of the value of the force in the tie rod will help in sizing the media to be used for static tests, for example.

© The Author(s) 2024

J.-J. H. Rincent, *Ground Anchors*, https://doi.org/10.1007/978-981-97-4414-5_3

CP	Dimensões	Secção	Limite de escoamento		Limite de resistência		Módulo elasticidade	Alongamento			Estricção	
	Ø									Lo		
	mm	mm²	kgf	MPa	kgf	MPa	GPa	Lo (mm)	%	L (mm)	Ø	%
1	8.01	50.39	7035	1370	8125	1580	206.61	50	11	55.48	6.67	31
2	8.00	50.27	7130	1390	8200	1600	198.03	50	12	55.86	6.36	37
3	8.00	50.27	7195	1400	8220	1600	198.80	50	12	55.94	6.12	41
4	8.00	50.27	6992	1360	8025	1570	199.22	50	11	55.66	6.22	40
5	7.95	49.64	6815	1350	7980	1580	200.25	50	10	55.15	6.58	31

Fig. 3.2 Example of direct traction results

Fig. 3.3 Sampling after
rupture

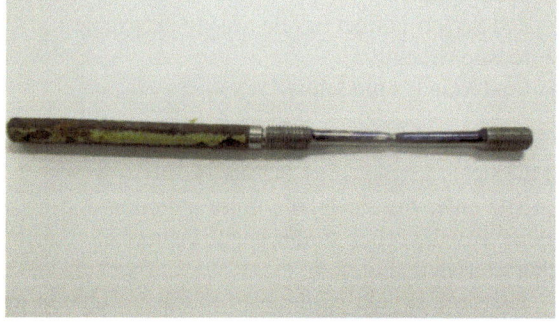

Circulaire	N°80 du 1/10/63	N°80 du 1/10/63	N°68 du 27/12/65	N°36 du 14/6/67	N°36 du 14/6/67	N°70-53 du 28/4/70
Usine	Bourg	Tréfimétaux	Bekaert	Sainte-Colombe	Providence	Arbed
Appellation	B 8-1	TH 7-1	7 BLS	7 ES	7 CFR 11	FG 8
Traitement acier	écroui vieilli	écroui vieilli	stress relieved	Patenté tréfilé stabilisé	Tréfilé et revenu	Patenté tréfilé revenu
Diamètre en mm	8	7	7	7	7	8
R_G en hectobars		132	167	157		162
T_G en hectobars	131	132	147	137	145	142
A_G en %	1,5				2	2
G à 120 h en %	6	NR	7	NR	2	6
G à 1000 h en %	8	10	10	4	3	8

Fig. 3.4 Characteristics of old steels

3.111. Traction de service

1. Aciers pour précontrainte

Si T_p désigne la force de traction correspondant à la limite d'élasticité conventionnelle de l'armature du tirant, la traction de service T_S doit être au plus égale aux valeurs suivantes :

- $T_S < 0,75\ T_p$ pour les tirants provisoires d'une durée d'utilisation ne dépassant pas dix-huit mois.
- $T_S < 0,60\ T_p$ pour les tirants permanents et les tirants provisoires devant jouer leur rôle pendant plus de dix-huit mois.

COMMENTAIRE

Ces valeurs sont fixées en fonction des risques de corrosion sous tension (voir article 4.03, p. 26).

Fig. 3.5 Extract from a catalog

Chapter 4
Length of the Tie Rod

The following figure visualizes the value Δf frequency, distance between two peaks that leads to the calculation of the lengths of the tested elements, also appear the admittance and the dynamic stiffness at the origin of the curve (Fig. 4.1).

The example that is presented comes from tests carried out on tie rods made 30 years ago to stabilize road retaining walls in a state in southern Brazil.

All the 664 tie rods on 6 retaining walls were tested. Our example is a retaining wall of 28 tie rods (Figs. 4.2 and 4.3).

Tie rods P1 Panel 1 Tie rod 2—P1 7—P2 2—P2 4—P3 4—P3 5 were not tested for several reasons, for example: plate partly in contact with the wall or deterioration of the tie rod head.

The results presented in the following table gather the total length of the ground anchor the free-length and the calculated sealed length (Fig. 4.4).

These results are compared to the design (Fig. 4.5).

Here after the curves obtained during the tests used to calculate total lengths and free lengths, see the curves in the following figures test P1 2C panel 1 tie rod 2 test C.

On each tie rod 8 tests are performed (Figs. 4.6 and 4.7).

The results obtained show that the tie rods produced correspond with the plans in terms of length.

It is necessary to draw attention to the fact that when calculating the length of the tie rods, the velocity of wave propagation in the tie rods is very often greater than 4000 m/s.

For example, the table below, taken from a test report, shows that the hypothesized retained velocities are 4500m/s and 5100m/s taking into account the site information (Fig. 4.8).

© The Author(s) 2024
J.-J. H. Rincent, *Ground Anchors*, https://doi.org/10.1007/978-981-97-4414-5_4

Fig. 4.1 Curve V/F function
of frequency

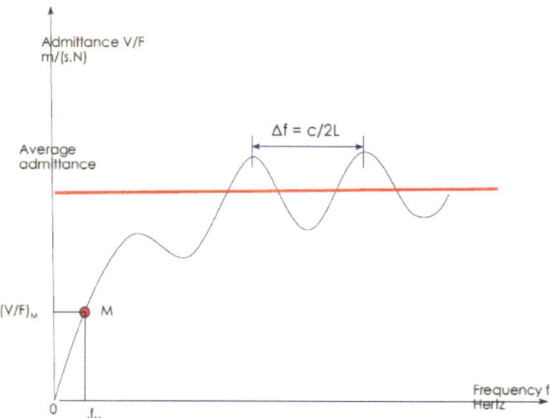

Fig. 4.2 Nondestructive test

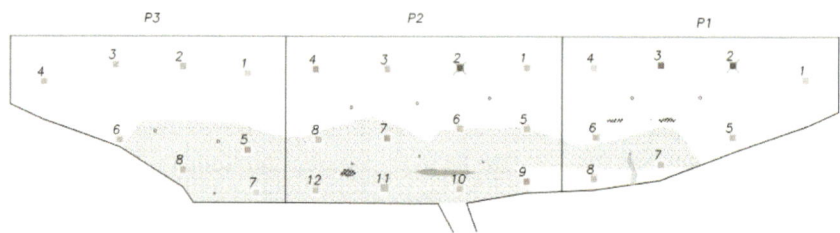

Fig. 4.3 Front view of the retaining wall

												average value meters	design value m
Line 1	**Total length**	18,2	18,2	17,9		19		20,1	19,4	19,1	18,1		
	free-length	11,3	10,6	11,5		10,4		10,2	10,6	10,6	10	10,7m	10m
	sealed-length	6,9	7,6	6,4		8,6		9,9	8,8	8,5	8,1	8,1m	8m
Line 2	**Total length**	18,6	14,7	15,6	16,1	17	16,7	16,8	16,3	14,7	19,3		
	free-length	10,8	5,6	8,8	8,5	8,8	8,5	9,6	8,9	6,3	10,8	8,66m	8m
	sealed-length	7,8	9,1	6,8	7,6	8,2	8,2	7,2	7,4	8,4	8,5	7,92m	8m
Line 3	**Total length**		15,6	15,1	15,6	14,6	15,3	15,4	14,2				
	free-length		6,4	6,2	6	5,9	6,7	6,5	6,7			6,34m	6 m
	sealed-length		9,2	8,9	9,6	8,7	8,6	8,9	7,5			8,77m	8m

Fig. 4.4 Results table

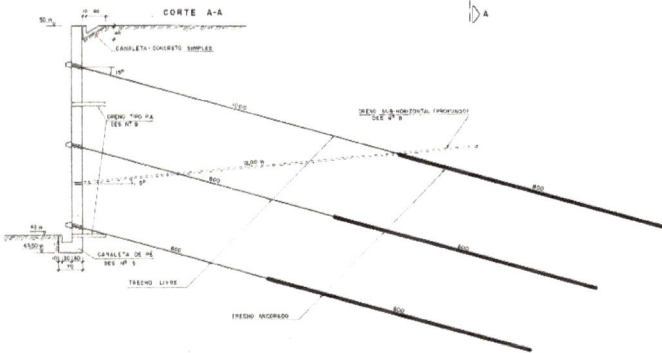

Fig. 4.5 Tie-rod execution drawing

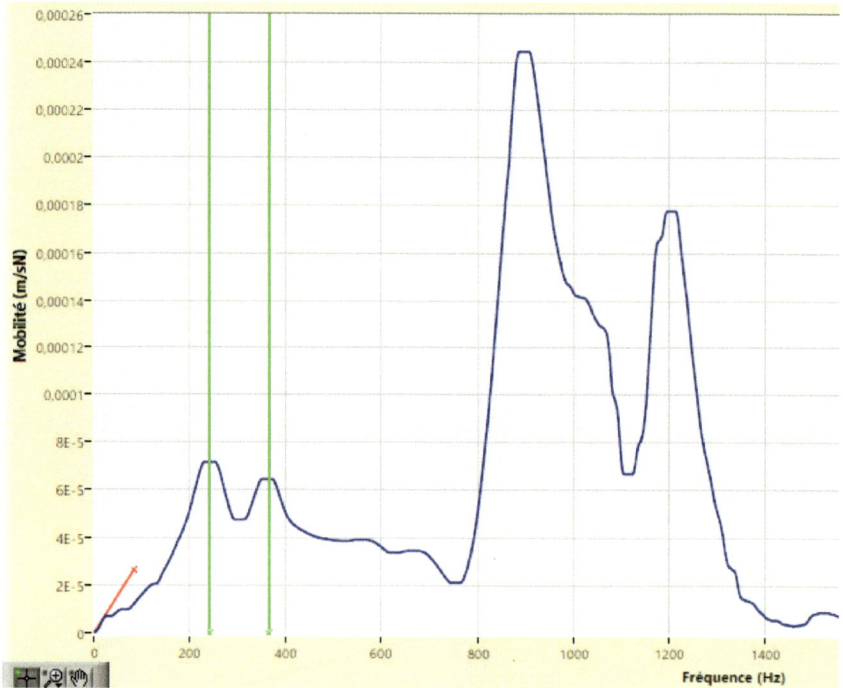

Fig. 4.6 Total length 18,1 m

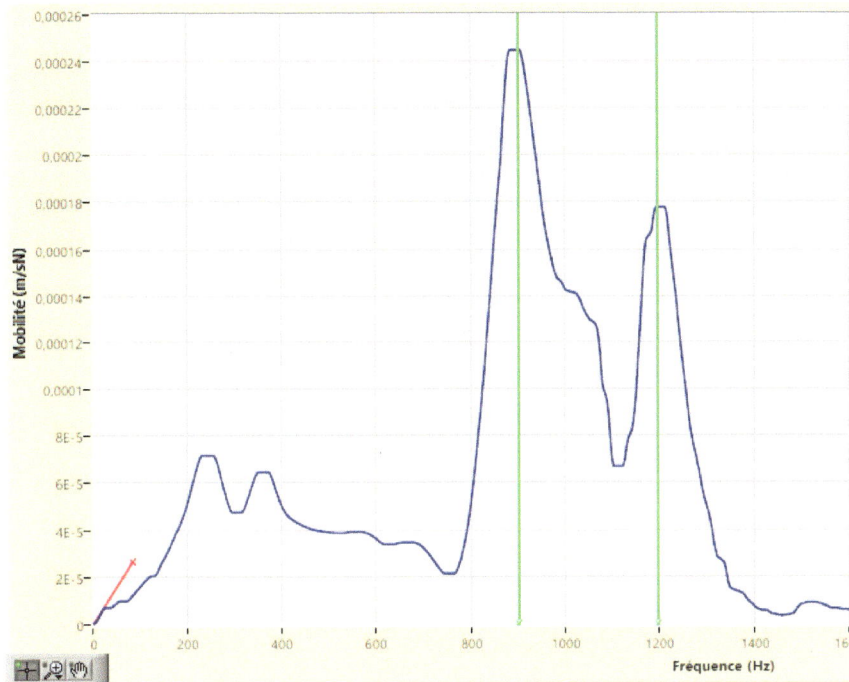

Fig. 4.7 Free-length 10 m

Painel N°	Tirante N°	Comprimentos calculados			
		Comprimentos Livres		Comprimentos Totais	
		4500 m/s	5100 m/s	4500 m/s	5100 m/s
	T3125	8,2	9,3	18,2	20,6
	T3126	5,8	6,5	26,0	29,5
	T3127	6,0	6,8	27,1	30,7
	T3128	8,2	9,3	27,2	30,8
	T3130	5,7	6,5	24,0	27,1
	T3132	6,2	7,0	22,7	25,7
	T3137	7,6	8,6	16,0	18,2
	T3138	6,5	7,4	18,9	21,4

Fig. 4.8 Table for calculating the lengths

Chapter 5
Dynamic Stiffness—Force

After discovering the principles linking dynamic stiffness to static stiffness and internal force in the tie bar, an international patent has been registered (Fig. 5.1).

Calibration.

Establishing the relationship between dynamic stiffness and tensile force in the tie rod that needs to eliminate the dynamic stiffness value for example of the retaining wall and other elements that can modify the total value of the stiffness (Fig. 5.2).

The test results obtained in Fig. 29, for example, led to the following analyses that relate the square root of dynamic stiffness to tensile force (Fig. 5.3).

The following figure collects a portion of the tension force results for a wall with 590 ties rod (Fig. 5.4).

Establishing the relationship of dynamic stiffness and tensile force in the tie that needs to eliminate the value of dynamic stiffness for example for the retaining wall and other elements that can modify the stiffness. The wall with beams is a very different configuration with a high inertia (Figs. 5.5 and 5.6).

This method cannot be used, for example, to measure force of a pre-stressed beam cable or a bridge deck. The only element that will be measured is the inertia of the beam or deck.

A study of a quay wall of a canal of a nuclear power plant shows that one tie, no. 2, was broken. The calculated average value of the internal force was 53 tons (Fig. 5.7).

© The Author(s) 2024
J.-J. H. Rincent, *Ground Anchors*, https://doi.org/10.1007/978-981-97-4414-5_5

(12) DEMANDE INTERNATIONALE PUBLIÉE EN VERTU DU TRAITÉ DE COOPÉRATION
EN MATIÈRE DE BREVETS (PCT)

(19) Organisation Mondiale de la Propriété
Intellectuelle
Bureau international

(43) Date de la publication internationale
2 février 2006 (02.02.2006)

PCT

(10) Numéro de publication internationale
WO 2006/010830 A1

(51) Classification internationale des brevets[7] : G01L 1/10,
G01N 3/34

(21) Numéro de la demande internationale :
PCT/FR2005/001597

(22) Date de dépôt international : 24 juin 2005 (24.06.2005)

(25) Langue de dépôt :   français

(26) Langue de publication :   français

(30) Données relatives à la priorité :
0406969   25 juin 2004 (25.06.2004)   FR

(71) Déposant (pour tous les États désignés sauf US) : RIN-
CENT BTP SERVICES [FR/FR]; 39 Rue Michel Ange,
Parc Elysée, F-91026 EVRY Cedex (FR).

(72) Inventeur; et

(75) Inventeur/Déposant (pour US seulement) : RINCENT,
Jean-Jacques [FR/FR]; c/o RINCENT BTP Services, 39

Rue Michel Ange - Parc Elysée, F-91026 EVRY Cedex
(FR).

(74) Mandataires : INTES, Didier etc.; Cabinet Beau de
Loménie, 158 Rue de l'Université, F-75340 PARIS (FR).

(81) États désignés (sauf indication contraire, pour tout titre de
protection nationale disponible) : AE, AG, AL, AM, AT,
AU, AZ, BA, BB, BG, BR, BW, BY, BZ, CA, CH, CN, CO,
CR, CU, CZ, DE, DK, DM, DZ, EC, EE, EG, ES, FI, GB,
GD, GE, GH, GM, HR, HU, ID, IL, IN, IS, JP, KE, KG,
KM, KP, KR, KZ, LC, LK, LR, LS, LT, LU, LV, MA, MD,
MG, MK, MN, MW, MX, MZ, NA, NG, NI, NO, NZ, OM,
PG, PH, PL, PT, RO, RU, SC, SD, SE, SG, SK, SL, SM,
SY, TJ, TM, TN, TR, TT, TZ, UA, UG, US, UZ, VC, VN,
YU, ZA, ZM, ZW.

(84) États désignés (sauf indication contraire, pour tout titre
de protection régionale disponible) : ARIPO (BW, GH,

[Suite sur la page suivante]

Fig. 5.1 Patent extract

Fig. 5.2 Simultaneity of
static and dynamic tests.
Source Rincent
BTP—France

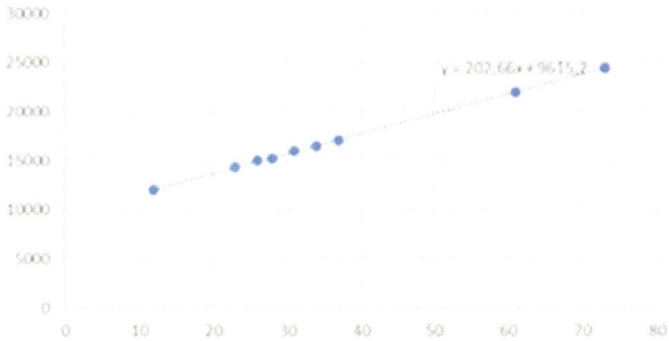

Fig. 5.3 Square root of dynamic stiffness as a function of force

7,3	4,7	6,5	2,4	9,4
12,5	9,7	3,3	9,1	13,7
9,9	11,0	13,4	5,8	10,5
18,9	13,9	15,0	4,1	12,6
10,1	14,6	12,7	12,4	7,3
12,0	16,0	14,9	6,1	12,6
15,8	19,0	18,1	17,7	17,0
24,5	19,0	20,8	21,3	21,2
19,0	16,9	18,0	16,7	19,5

Fig. 5.4 Force results table in tons

Fig. 5.5 Site overview.
Source Rincent
BTP—Recife

Fig. 5.6 Detailed view of
the concrete beams. *Source*
Rincent BTP—Recife

tie-rod number	1	2	3	4	5	6	7
Force (t)	24	broken	47	82	81	67	57

8	9	10	11	12	13	14	15
45	42	53	56	55	45	45	69

Fig. 5.7 Effort distribution force (t) Tie rod number

Chapter 6
Diameter of the Tie-Rod with Its Cement Grout

See Fig. 6.1.

First example of calculations from in situ tests (Fig. 6.2).

The formula of the NF standard P94 160–4.

6.1 $V/F = 1 / \rho_b \, V \, B \, a$

Concrete volume mass ρ_b in kg/m3.

Circular cross section (m^2).

Mobility 1,5 10–6 m/sN.

Concrete bulk mass—hypothetical 2500 kg/m3.

Wave velocity inside the tie, hypothesis 4500m/s.

Right section of the reinforcement A.

$A = 0{,}06$m^2

$D = 0{,}27$m.

See Fig. 6.3.

Second example (Fig. 6.4).

Mobility 4,46 10–5 m/sN.

Volumetric mass of concrete and bar 2500 kg/m3.

Wave celerity inside the tie rod 4500m/s.

$A = 2$ E-3 m^2

This result is normal, because these bars are sealed in the rock.

$D = 0.05$m (Figs. 6.5 and 6.6).

© The Author(s) 2024
J.-J. H. Rincent, *Ground Anchors*, https://doi.org/10.1007/978-981-97-4414-5_6

Fig. 6.1 Pulled out of the tie-rod with its grout. *Source* Thesis and Photo Thiago Benjamin Porto-P49

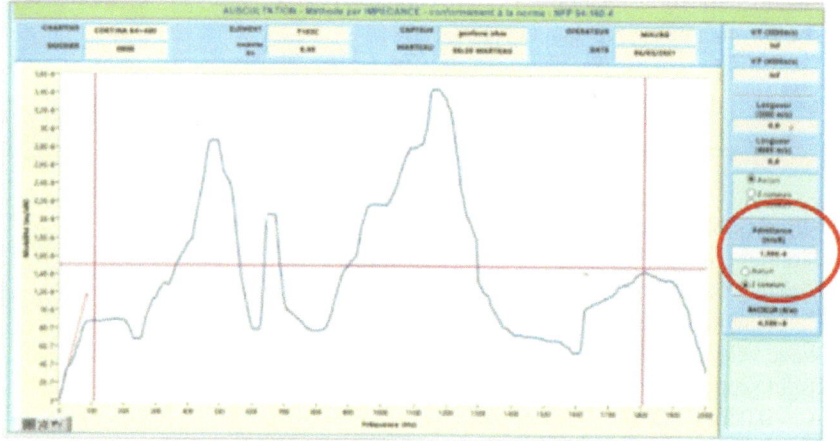

Fig. 6.2 Mobility measurement

Fig. 6.3 Calculation made for this type of anchoring with strands. *Source* Rincent BTP—Recife

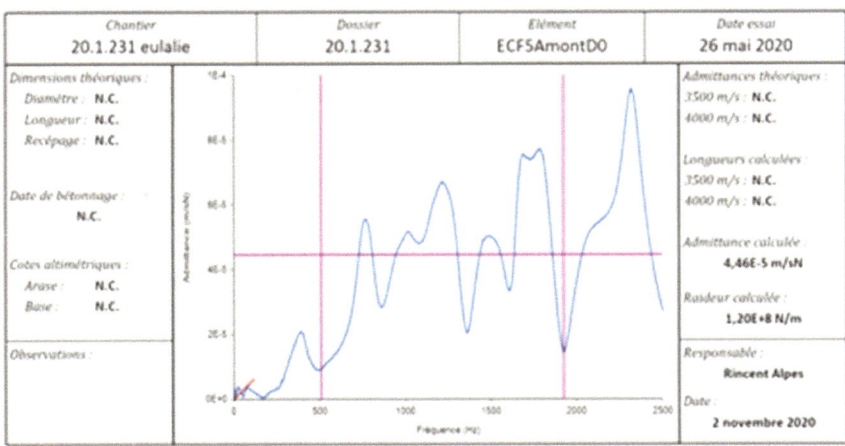

Fig. 6.4 Mechanical admittance

Fig. 6.5 Cliff reinforcement bars. *Source* Rincent BTP—France

Fig. 6.6 Cliff reinforcement static tests. *Source* Rincent BTP—France

Chapter 7
Horizontal Anchors

Embankments contained by two tensioned retaining walls are an opportunity to control our measurements. Tests are performed at both ends of the tie rods and the results are compared (Fig. 7.1).

The dynamic stiffness is calculated from the eight mechanical impedance tests performed on the tie rod head.

Two tie rods were tested on both sides of the embankment (Figs. 7.2 and 7.3).

First case

The values for the forces are as follows, 8.7 tons and 8.6 tons, i.e. 0.1-ton difference.

Second case

The force values are 6.9 tons and 7.1 tons or 0.2ton difference.

The measurement uncertainty is less than 0.2 ton.

On another site, tests were performed with the following results.

The characteristic of the tie rods being formed by a central plate and two bars, 16 acquisitions were performed, 8 on the left bar, 8 on the right bar. Each of these acquisitions results in a dynamic stiffness value.

T1	Left	3,67E + 08	**8,4t**	Right	3,86E + 08	**8,7t**
T2	Left	4,43E + 08	**4,43t**	Right	4,40E + 08	**4,40t**
T3	Left	5,04E + 08	**10,3t**	Right	4,82E + 08	**10,0t**

The differentials of the internal force values are: 0.3 ton—0 ton and 0.3 ton.

Five tests were performed on horizontal ties the difference between the calculated forces on both sides of the tie backfill is less than 0.5 ton (Fig. 7.4).

© The Author(s) 2024   49
J.-J. H. Rincent, *Ground Anchors*, https://doi.org/10.1007/978-981-97-4414-5_7

Fig. 7.1 Horizontal ties across an embankment. *Source* Rincent BTP—Recife

114	5.32E+08		112	5.65E+08	
	5.36E+08			5.11E+08	
	5.63E+08			5.34E+08	
	5.03E+08			5.43E+16	
	5.58E+08			5.63E+08	
	5.42E+08			5.43E+08	
	5.21E+08			5.35E+08	
	5.34E+08	5.34E+08 8.6t		5.33E+08	5.42E+08 8.7t

102	3.78E+08		103	4.34E+08	
	3.95E+08			3.95E+08	
	4.19E+08			4.04E+08	
	3.80E+08			4.22E+08	
	3.82E+08			3.60E+08	
	3.69E+08			4.45E+08	
	3.98E+08			5.37E+08	
		3.90E+0			4.08 E+0
	4.04E+08	8 6.9t		3.62E+08	8 7.1t

Fig. 7.2 Results of stiffness (N/m) and force (t)

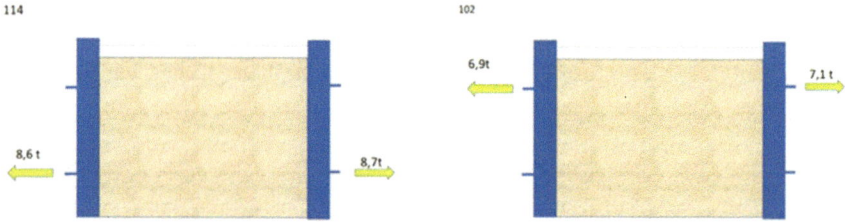

Fig. 7.3 Walls and anchors

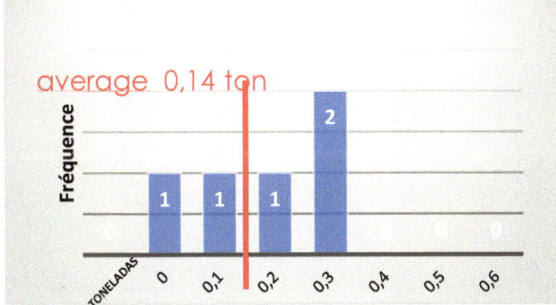

Fig. 7.4 Histogram of differences

Chapter 8
Earth Pressure

One of our diagnostics was performed on 590 tie rods positioned on 12 lines. The analysis of the force values for each line of tie rods shows a distribution of the thrust forces in accordance with the theory (Fig. 8.1).

For each line of tie rods, the force histogram was made with the calculation of the average force value per line.

Histograms of force values per tie line (Figs. 8.2 and 8.3).

A calculation of the thrust forces was carried out with the following assumptions.

$C = 1.5$ t/m^2 cohesion.

Phi $= 25°$—angle of internal friction.

and is compared to the results of the dynamic stiffness calculation (Fig. 8.4).

The force values calculated on the last three tie rows are influenced by the presence at the foot of the wall of a concrete structure that collects and evacuates runoff water (Fig. 8.5).

The preceding observations draw attention to the representativeness of the tests on the anchors. The tests carried out mainly at the foot of the wall for reasons of ease of access constitute a particular approach which in general is not representative of common cases.

© The Author(s) 2024
J.-J. H. Rincent, *Ground Anchors*, https://doi.org/10.1007/978-981-97-4414-5_8

Fig. 8.1 Photo of the wall. *Source* Rincent BTP—Recife

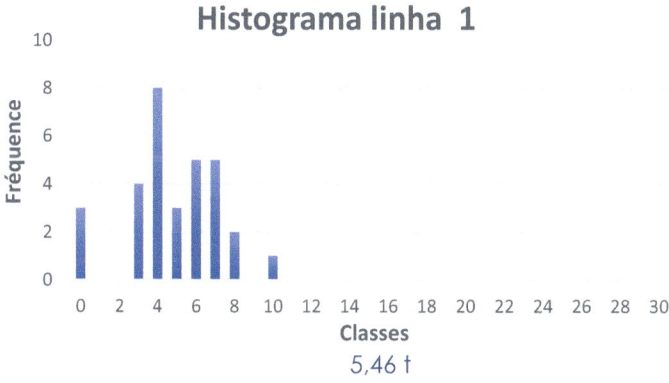

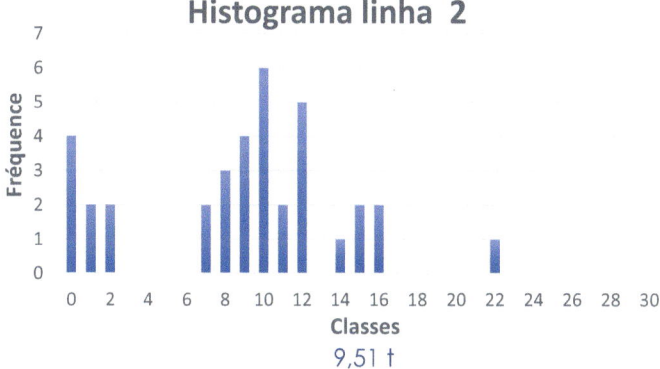

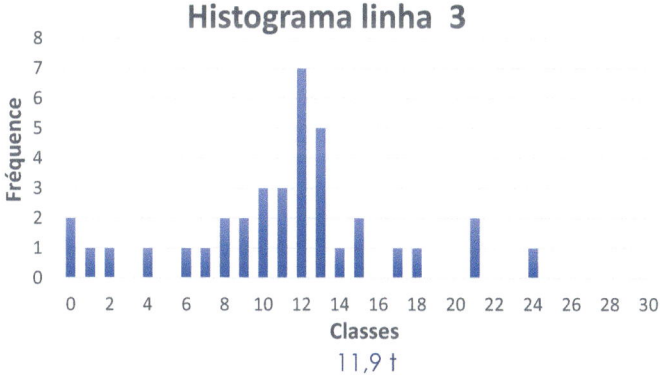

Fig. 8.2 Histograms of force values per tie rod line

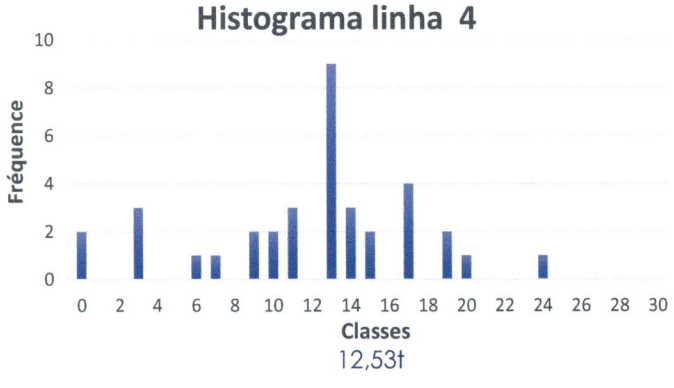

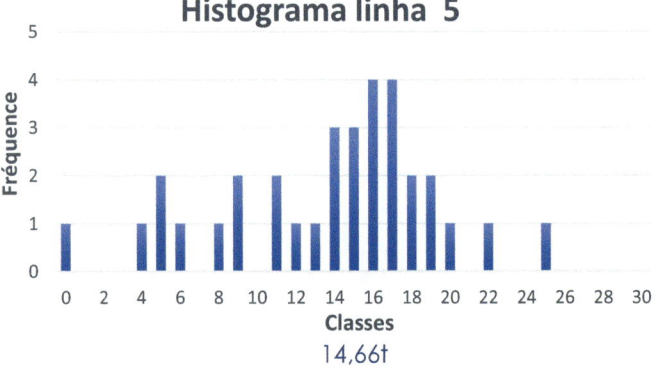

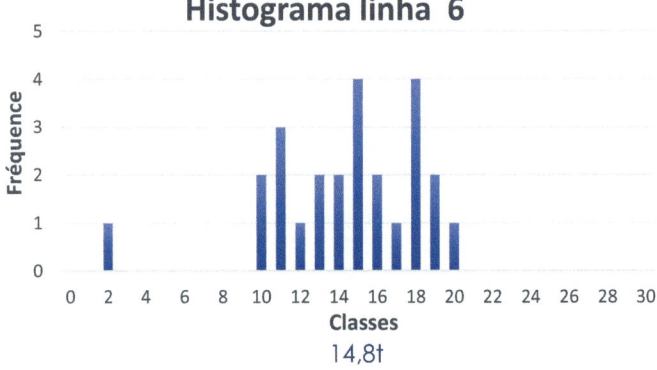

Fig. 8.2 (continued)

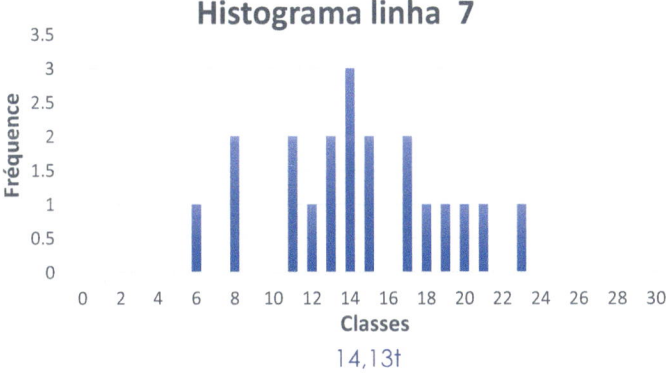

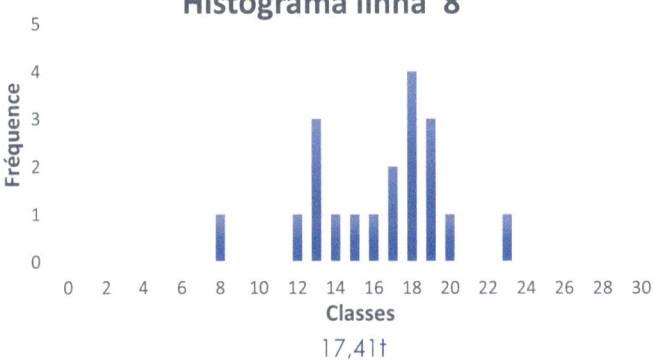

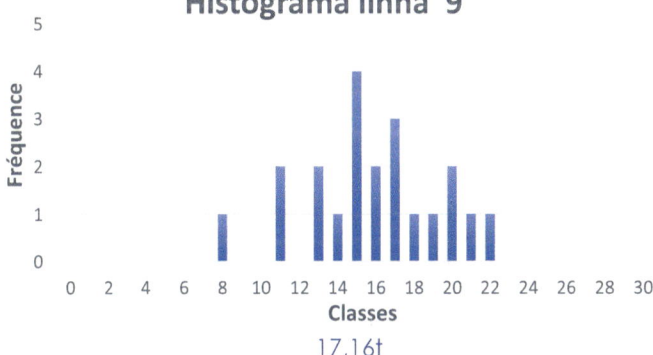

Fig. 8.2 (continued)

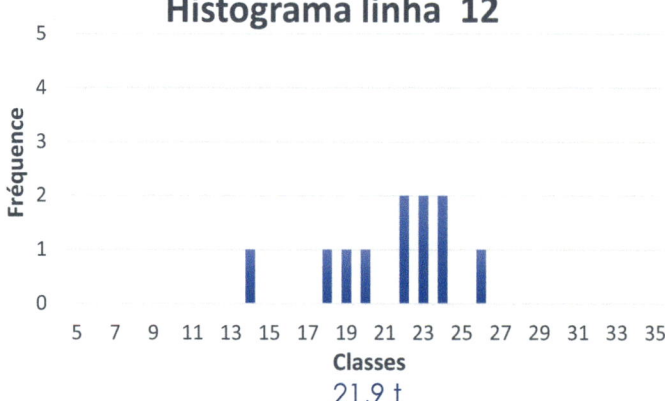

Fig. 8.2   (continued)

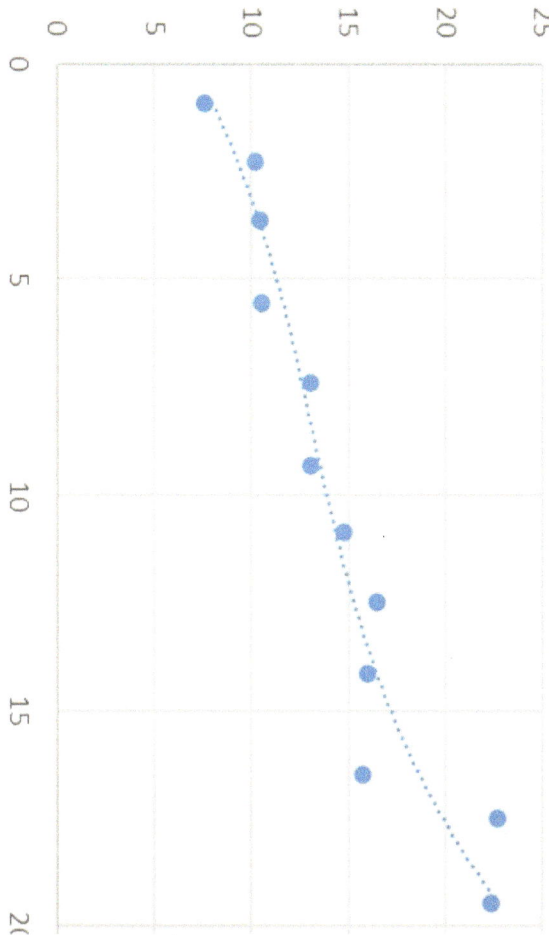

Fig. 8.3 Force (tons) as a function of wall height

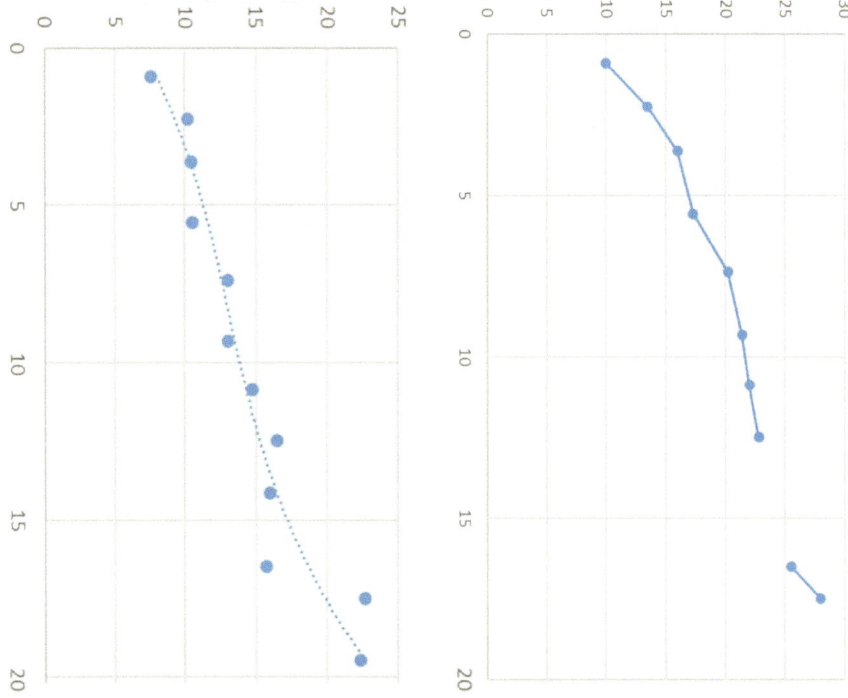

Fig. 8.4 Force curves as a function of wall height

Fig. 8.5 Concrete structure.
Source Rincent
BTP—Recife

Chapter 9
Static Tests

The objectives of the static tests are:

- to determine the internal force at the tie rod and to calibrate the dynamic method
- to calculate the static stiffness
- and simultaneously perform dynamic tests to calculate the dynamic stiffness
- to analyze the tie rod behavior with the aim of a tie rod re-tension.

The following graphical construction shows that the tensile force given by the operator is 3t higher than the one calculated graphically during unloading.

The operators who performed the static test determined the tension force of the tie rod during the loading phase, the value is 23 tons.

The graphical construction made with the elements of the unloading phase leads to a tensile force of 20 tons.

The calculation from the dynamic stiffness gives a result of 20.4 tons.

The determination of the tensile force during the increasing load phase is obtained when the metal plate lifts off the head of the tie rod.

That is, when the applied force is greater than the tension force plus the adhesion forces of the plate to the support.

When the tie rods are old, the force necessary to detach the metal plate located between the head of the tie rod and the wall is 1 to 2 tons higher than the tension force. This adhesion of the system on the wall is all the stronger that there is corrosion (Fig. 9.1).

When the tie rods were re-tensioned, the following test results were analyzed:

- Static tests
- Dynamic tests
- And re-tensioning operation.

The tensile strengths calculated from the static tests are generally 1.5 tons higher than the values calculated from the re-tensioning results.

© The Author(s) 2024

J.-J. H. Rincent, *Ground Anchors*, https://doi.org/10.1007/978-981-97-4414-5_9

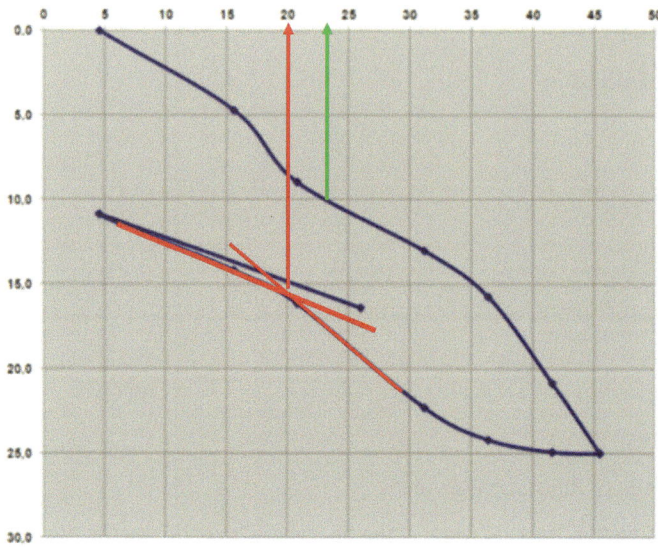

Fig. 9.1 Displacements (mm) as a function of tensile force in tons

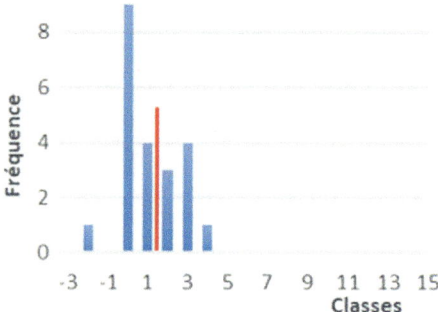

Fig. 9.2 Average 1,46t—Standard deviation 1,6t

Tensile strengths calculated with nondestructive tests are close to the values calculated with the prestressing results (Figs. 9.2 and 9.3).

Fig. 9.3 Static tests at height. *Source* Rincent BTP—Recife

Chapter 10
Interest of Non-destructive Tests

Static testing requires the use of heavy equipment that is difficult to handle at height.

Performing static tests also means running the risk of breaking the tie rod, especially in the case of old anchoring.

The installation of test and measurement equipment is difficult and time consuming, especially when the test is performed at height.

In the end, these tests are expensive and their number is very limited.

For non-destructive testing, the equipment is light and allows access at height. The test itself is of short duration.

Under these conditions, it is economically wise to limit static tests and to increase the number of non-destructive tests to improve the representativeness of the diagnosis on a tensioned wall (Fig. 10.1).

Fig. 10.1 Nondestructive test. *Source* Rincent BTP—Recife

J.-J. H. Rincent, *Ground Anchors*, https://doi.org/10.1007/978-981-97-4414-5_10

Chapter 11
Corrosion

The first operation consists in measuring the diameter of the steels to check their dimensions, their corrosion (Fig. 11.1).

Direct tensile tests on steel samples taken from the site are a prerequisite.

The other important issue with steels is corrosion of the tie rod. The Swiss standard SAI 267/1 (2013) defines the acceptability of a newly injected tie rod by establishing minimum resistivity values. For example, these values can be used as criteria for re-tensioning.

Extracts from the standard concerning electrical resistance measurements are attached below.

10.7.4.2 A tie rod, once injected and energized, shall have an electrical resistance RI 0.1 MΩ (Mega Ohm).

10.7.4.3 In case a certain percentage of tie rods not showing electrical resistance I is tolerated, e.g. 5–10%, an electrical resistance II (ERM II) measurement shall be carried out on the defective tie rods to prove that there is no galvanic contact between the head of the tie rod and the frames of the structure. The test shall be performed in accordance with SIA 267/1.

The electrical resistance between the head of the rod and the frames shall have an RII value of 100 Ohms (Figs. 11.2 and 11.3).

It is interesting to note that even tie rods installed over 40 years ago can meet new tie rod requirements. The quality of the initial construction is related to these results.

© The Author(s) 2024
J.-J. H. Rincent, *Ground Anchors*, https://doi.org/10.1007/978-981-97-4414-5_11

Fig. 11.1 Diameter measurement. *Source* Rincent BTP—Recife

Fig. 11.2 Measuring the
electrical insulation of tie
rods. *Source* Rincent
BTP—Recife

TIRANTE Nº T709	FIO 1	FIO 2	FIO 3	FIO 4	FIO 5	FIO 6	FIO 7	FIO 8
LEITURA 1 (MΩ)	7	6	6	6	6	6	6	6
LEITURA 2 (MΩ)	7	6	6	6	6	6	6	6
LEITURA 3 (MΩ)	7	6	6	6	6	6	6	6
TIRANTE Nº T703	FIO 1	FIO 2	FIO 3	FIO 4	FIO 5	FIO 6	FIO 7	FIO 8
LEITURA 1 (MΩ)	9	10	10	10	11	10	10	11
LEITURA 2 (MΩ)	9	10	10	10	10	10	10	10
LEITURA 3 (MΩ)	9	10	10	10	10	10	10	10

Fig. 11.3 Measurement results

Chapter 12
Load Cells

12.1 Example

A railway company has implemented a tension monitoring of 4 tie rods stabilizing an embankment. This monitoring has been done with load cells for more than 25 years.

One cell shows that the force decreases and the other three show that the tension force remains constant.

The concessionaire requested a non-destructive test, which we performed. The question concerned cell 1, which has a questionable behavior (Figs. 12.1 and 12.2).

From the dynamic tests, we calculated the dynamic stiffness of the tie rods.

Tie Rod 1 7,31E + 8 N/mTie Rod 2 7,03E + 8 N/m.

Tie Rod 3 6,58E + 8 N/m Tie Rod 4 6,19E + 8 N/m.

This leads to the following graph (Fig. 12.3).

Using the equation from the diagram in the previous figure it is possible to calculate the real tension force of tie rod 1 which is 473 kN (Fig. 12.4).

The results of the non-destructive tests show the reliability of the method and the different uses that can be made of it.

12.2 Load Cells

The cells used in the example in the previous chapter are ring cells that have been in service for over 25 years (Fig. 12.5).

The vibrating strings are used to know the value of the tension in different points of the same tie to know its operation.

To obtain the results it is necessary to excite the vibrating string with an electric current so that the vibrating string gives a vibratory response in relation to the tension force (Fig. 12.6).

© The Author(s) 2024
J.-J. H. Rincent, *Ground Anchors*, https://doi.org/10.1007/978-981-97-4414-5_12

Fig. 12.1 Load cells. *Source*
Rincent BTP—France

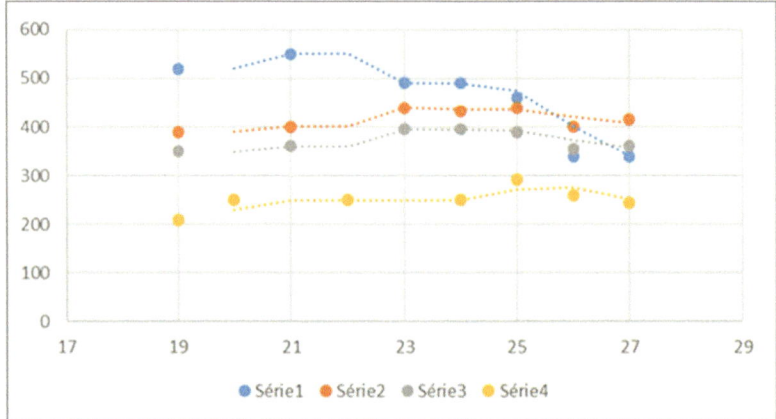

Fig. 12.2 Forces measured by the load cells

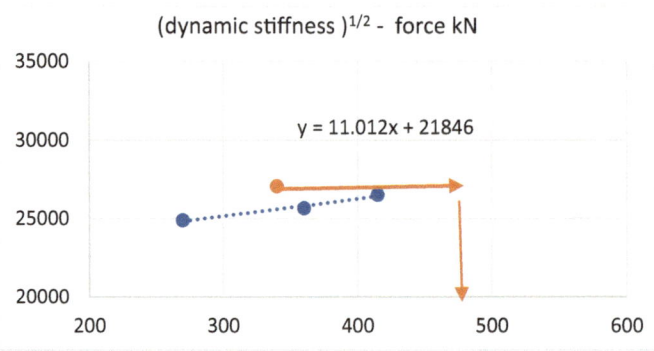

Fig. 12.3 Square root of stiffness versus force (kN)

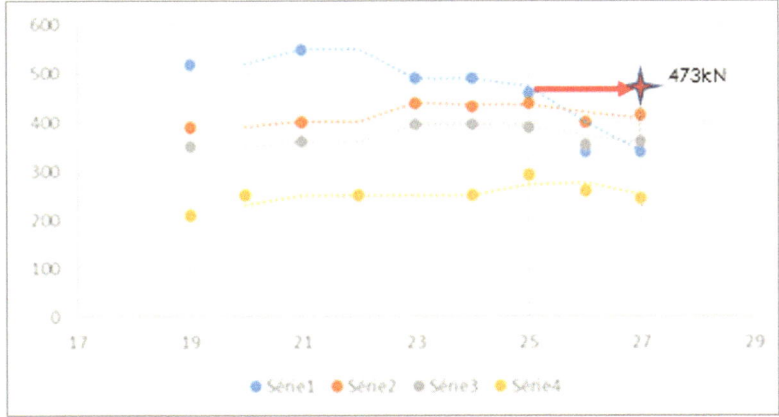

Fig. 12.4   Real value of the force—gauge 1

Fig. 12.5   Ring cells. *Source* Glötz Company

Fig. 12.6   Vibrating string.
Source Rincent BTP France

Chapter 13
The Fatigue Phenomenon

The fatigue phenomenon is well known in mechanics, but less so in civil engineering.

A simplified presentation could be the following: a steel bar that breaks in tension for a force of one ton will break under a cyclic force of 100 kg performed 10E6 or 10E7 times.

During a static tensile test, a video made for the operators showed the effects of the load transmitted by the axles during the passage of a train.

For each axle, measuring devices recorded a load-induced deformation (Fig. 13.1).

Depending on the position of the applied load relative to the wall, the second row of tie rods generally has the lowest force values, this is especially true for transmitted dynamic loads. These overloads are usually applied for short periods of time (Figs. 13.2, 13.3 and 13.4).

The behavior changes at 32.5 tons and above, and the tie rod in the fourth row does not support a higher load.

The main difference is that the tie rod in Fig. 96 only receives static loads from the ground, whereas the tie rod in Fig. 97 received dynamic stresses from the passing trains in addition to the ground thrust.

The induced mechanical fatigue is a very important factor, and the behavior of the tie rod under the loads cannot resist a strong mechanical action.

For an assumption of 10 trains of 100 wagons per day with 4-axles for 200 days per year for 40 years, the result is 3,2E7 dynamic overloads.

This calculation, which is an evaluation, must be compared with the Wölher curve which is used in mechanics courses on fatigue (Fig. 13.5).

The displacement versus load curve of a railway train shows that the induced vertical deformations are mil8imetric (Fig. 13.6).

Harbors are another location where cyclic loads are applied daily on anchors.

Deep-water tie rod tests are often conducted during high tides, as the low tide level allows access to the usually submerged tie rod heads.

At this location, the tidal range was 6 m.

© The Author(s) 2024
J.-J. H. Rincent, *Ground Anchors*, https://doi.org/10.1007/978-981-97-4414-5_13

Fig. 13.1 Transmission of the load. *Source* Rincent BTP—Recife

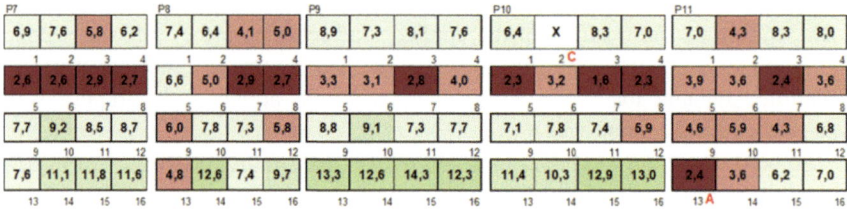

P7

6,9	7,6	5,8	6,2
1	2	3	4
2,6	2,6	2,9	2,7
5	6	7	8
7,7	9,2	8,5	8,7
9	10	11	12
7,6	11,1	11,8	11,6
13	14	15	16

P8

7,4	6,4	4,1	5,0
1	2	3	4
6,6	5,0	2,9	2,7
5	6	7	8
6,0	7,8	7,3	5,8
9	10	11	12
4,8	12,6	7,4	9,7
13	14	15	16

P9

8,9	7,3	8,1	7,6
1	2	3	4
3,3	3,1	2,8	4,0
5	6	7	8
8,8	9,1	7,3	7,7
9	10	11	12
13,3	12,6	14,3	12,3
13	14	15	16

P10

6,4	X	8,3	7,0
1	2 C	3	4
2,3	3,2	1,6	2,3
5	6	7	8
7,1	7,8	7,4	5,9
9	10	11	12
11,4	10,3	12,9	13,0
13	14	15	16

P11

7,0	4,3	8,3	8,0
1	2	3	4
3,9	3,6	2,4	3,6
5	6	7	8
4,6	5,9	4,3	6,8
9	10	11	12
2,4	3,6	6,2	7,0
13 A	14	15	16

Fig. 13.2 Measured tension forces (tons). Force in tons deformations in mm

Tests carried out on the same day at different times on the same anchor show a difference in force of 15 tons for an average force of 100 tons. This phenomenon, related to the thrust of the water on the sheet pile curtain, is well known. These cyclic loads of varying amplitudes are part of the anchors loads.

Force in tons deformations in mm

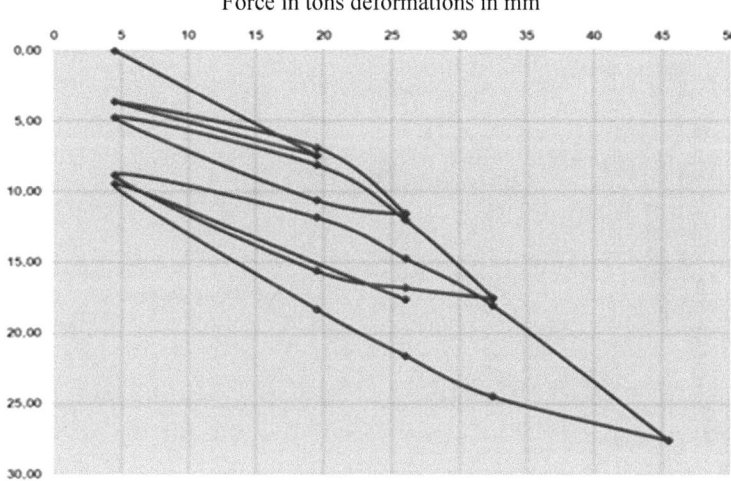

Fig. 13.3 Static test on a 40 years old tie rod not subjected to dynamic overload. Force in tons deformations in mm

Force in tons deformations in mm

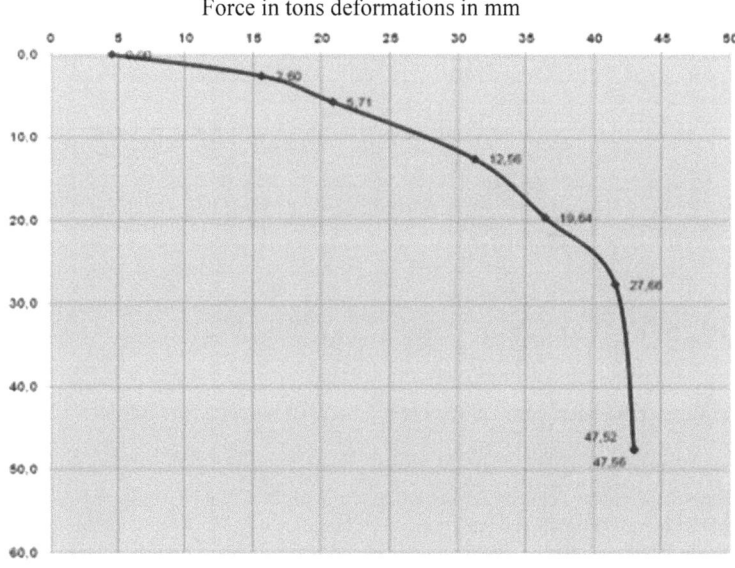

Fig. 13.4 Static test of a tie rod receiving a dynamic overload

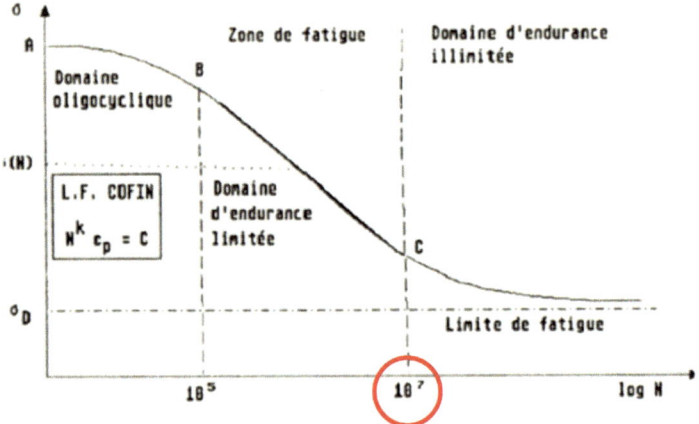

Fig. 13.5 Wölher curve

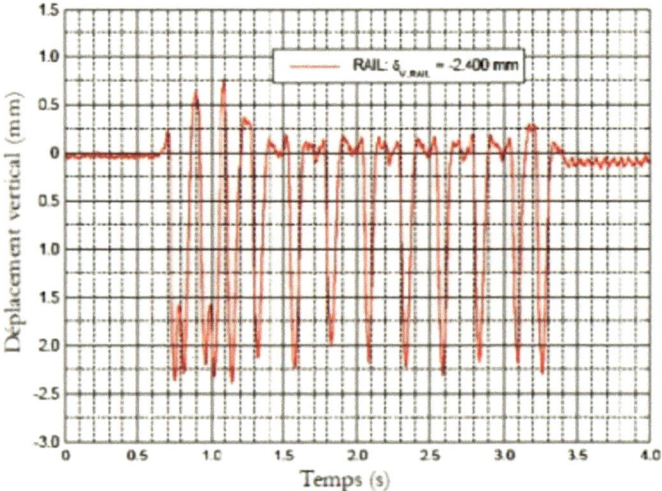

Fig. 13.6 Vertical displacements, railway convoy. *Source* Laurent Soyez Thesis

Chapter 14
Loss of Tension

Prestressed tie rods lose a percentage of their tensile force with time.

This is a general phenomenon, even for tie rods in dams where the tie rods are anchored in concrete on rock. The organizations responsible for the management of dams provide important information for many years.

These tension losses lead to a decrease in the safety factor with respect to the stability of retaining walls.

Diagnosis and maintenance of tie rods are necessary to guarantee the durability of the structures. The managers of retaining structures are aware of the phenomenon of loss of load in the tie rods and are beginning to request periodic dynamic tests to check the evolution of the tensile strength of the tie rods.

Examples
27 Tie rods
Initial tension 25 tons
After 24 years
Average value 19,9 tons
Loss of the tension force by 0.96% per year
15 tie rods
Initial tension 15 tons
After 24 years
Average value 7,78 tons
Loss of tension de 4,28% per year
6 tie rods
Initial tension 27 tons
After 5 years 18,5 tons
Loss of tension 7,28% per year
5 tie rods
Initial tension 32 tons
After 5 years 16,2 tons
Loss of tension 12,73% per year

© The Author(s) 2024
J.-J. H. Rincent, *Ground Anchors*, https://doi.org/10.1007/978-981-97-4414-5_14

Fig. 14.1 Loss of tension

Wall number	Tested Elements	Force tons
1	27	6,76
2	109	6,6
3	30	6,3
4	25	8,1
5	104	7,4
6	160	7,7
7	270	6,6
	725	
		7,05t

Fig. 14.2 Residual force per line of tie rods

	Average Force tons	residual force
Line 1	6,1	17,40%
Line 2	7,7	22,00%
Line 3	7,25	20,70%
Line 4	9,6	27,40%
Line 5	13,16	37,60%

We carried out tests on 7 retaining walls on 725 tie rods with steel bars as reinforcement. The initial tension value was 35 tons. Carried out 30 years ago, the average tension value is now 7 tons, i.e. an average loss of 5.5 % per year (Fig. 14.1).

For the wall n°7 the residual force for each tie rod line is (Fig. 14.2):

The loss of tension in the tie rods leads to approaching the limit equilibrium of the wall's stability with impermissible displacements (Fig. 14.3).

Another work, on 19 retaining walls and with 1173 nondestructive tests, have a single retaining wall where 590 tie rods were tested, stress loss analyses were performed for each of the 12 tie lines (Fig. 14.4).

Lines 1 à 4 lose 75% à 77%.

Lines 5 et 6 lose more than 60% and less than 70%.

Lines 8 à 10 lose 51% to 53%.

The last two lines lost 33%.

The French nuclear safety organization refers to the problem of load loss in the tie rods.

Regarding losses due to relaxation of the steel, Electricité de France has maintained a relaxation rate of 4% over 1000 h.

This value depends on the nature of the steel used and its heat treatment. For steels that do not undergo any special treatment to reduce relaxation, a relaxation at 1000 h of 8 to 9% is generally retained.

Fig. 14.3 Non admissible displacement. *Source* Rincent BTP—Recife

line	tested anchors number	loss of tension	
1	75	77,00%	
2	72	76,00%	
3	71	75,00%	
4	68	75,00%	
5	65	69,00%	
6	57	61,00%	
7	50	56,00%	
8	45	51,00%	
9	42	52,00%	
10	20	53,00%	
11	14	33,00%	
12	11	33,00%	
	590		

Fig. 14.4 Loss of tension per line de tie rods

Chapter 15
Re-tensioning

The analysis of the dynamic tests and static tests allows us to identify the tie rods that can be re-tensioned.

The static tests allow to know the deformations of the tie rods under the selected re-tensioning loads. The deformation generated to reach the new load will be stabilized by the introduction of metal plates of different thicknesses.

The procedure described below consists in inserting metal plates of different thicknesses between the wall and the head of the tie bar to tighten the tie bar to the value defined by the engineering department.

Example of procedure (Fig. 15.1).

In order to perform this re-tensioning, it is necessary to design and implement a traction device that allows this operation.

All re-tensioned tie rods were subjected to electrical isolation measurements, which was a first requirement for this operation (Fig. 15.2).

The analysis of the results focuses on the 24 10-strand ties, for which there are more results. The chosen tensioning force is 26 tons.

We consider as correct the tension force provided by the thickness of the plates inserted between the support plate and the wall (Fig. 15.3).

The figure below shows the difference between the tensile force values from the static tests and those deduced from re-tensioning the tie rods to 26 tons (Fig. 15.4).

Average + 0,84.

Standard deviation 2,70.

The force values above 1, 2, 3 and 4 tons may be related to corrosion and the additional force required to lift the tie rods off the plates and wall.

For the 7, 8 and 9 tons, the results are very high and the values are not in conformity.

The figure below shows the difference between the values of the tension forces from the dynamic tests and those deduced from the re-tensioning of the tie rods for 26 tons (Fig. 15.5).

Four values are abnormally low and can be attributed to a non-conformity of the acquisition, for example the bad fixing of the sensor on the head of the tie rod.

© The Author(s) 2024
J.-J. H. Rincent, *Ground Anchors*, https://doi.org/10.1007/978-981-97-4414-5_15

Fig. 15.1 Re-tensioning procedure

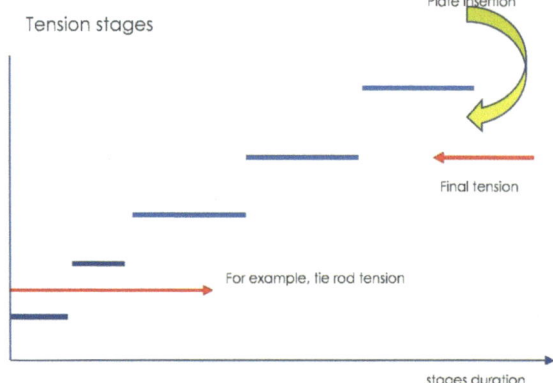

Fig. 15.2 Re-tensioning. *Source* Rincent BTP—Recife

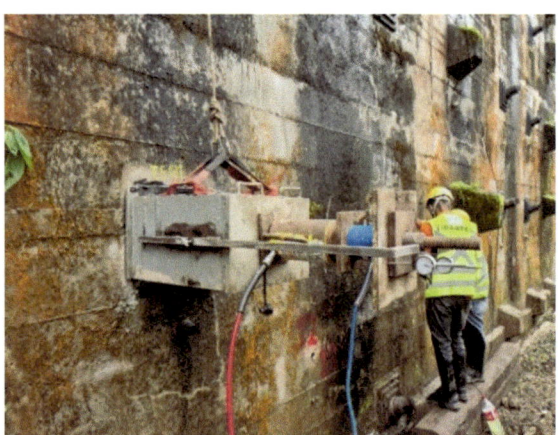

Excluding these values:

Average -0,85 t

Standard deviation 2,57

The results are close to the static tests in absolute values.

In order to reintroduce the value of 26 tons in the tie rod, it is necessary to reach a higher level of force to be able to introduce the metal plates.

This operation may disturb the behavior of the tie rod. Finally, it should be noted that the calculated forces from the static test are slightly higher than the actual force internal to the tie rod and for the dynamic tests the calculated force is slightly lower.

Re-tensioning and balance force

The following example shows the evolution of the force in a re-tensioned tie. The initial force was 30 tons, after 27 years, the internal force in the tie bar has stabilized at 25 tons (Fig. 15.6).

405	406 8,1	407	408	409	410	411	412
418	419 10	420	421	422	423	424	425 20
431 10	432 13	433	434	435	436	437	438 12
444 7,5	445	446	447	448	449	450 1,3	451 0
457 10	458 22	459	460	461	462	463	464 16
470	471	472	473	474	475	476	477
483	484	485	486	487	488	489	490
496	497	498	499	4100	4101	4102	4103
4109	4110	4111	4112	4113	4114	4115	4116
4122	4123	4124	4125	4126	4127	4128	4129 26
4135	4136	4137	4138	4139	4140	4141	4142 26
	4144	4145	4146	4147	4148	4149	4150
501	502	503	504	505	506	507	508
514	515 12	516	517	518 queb	519	520	521
527	528 7	529	530 10	531 12	532	533	534
540	541 0,3	542	543 26	544 0	545	546	547
553	554 12	555	556 22	557	558	559	560
566	567	568	569	570	571	572	573
579	580	581	582	583	584	585	586
592	593 22	594	595	596	597	598	599
5105	5106	5107	5108	5109	5110	5111	5112
5118	5119	5120	5121 26	5122	5123	5124	5125
5131 19	5132 22	5133	5134 12	5135	5136	5137	5138
5144	5145	5146	5147	5148	5149		

Fig. 15.3 Number of tie rods and calculated load values in tons

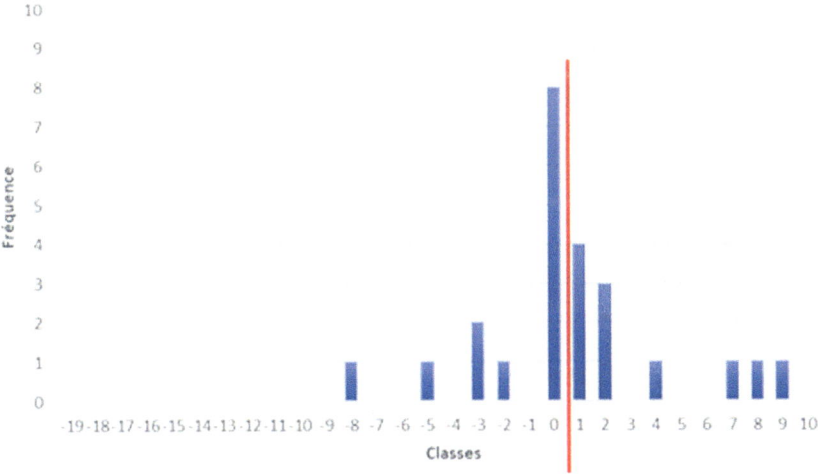

Fig. 15.4 Difference in force values from static and re-tensioning tests

Fig. 15.5   Force difference between dynamic and re-tensioning tests

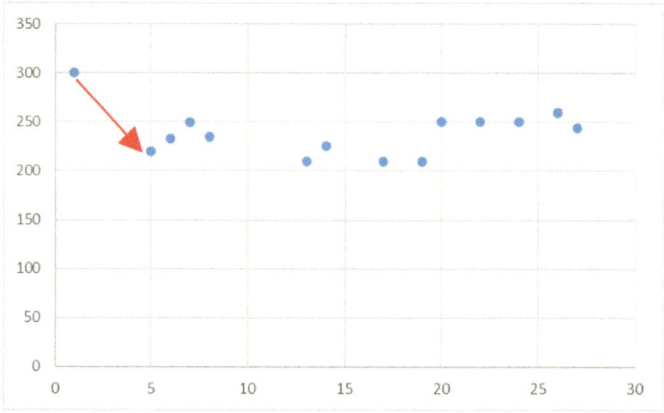

Fig. 15.6   Values of the tension force (kN) according to the years

Then a re-tensioning of the tie rod to 40 tons was decided. After three months, the force decreases to 27 tons and after five months, the force returns to the initial values. The equilibrium situation before the re-tensioning is a reality of the site (Fig. 15.7).

Protection of the head of the tie rod

Examples of tie rod head protection are given in TA 2020, published year 2020 (21).

These devices are completely filled with anti-corrosion agent, attached to the support plate.

This protection allows easy access to the tie rod head for re-tensioning (Fig. 15.8).

Note that non-destructive testing can be performed with protection. Corrosion protection products, waxes and petroleum-based greases are commonly used.

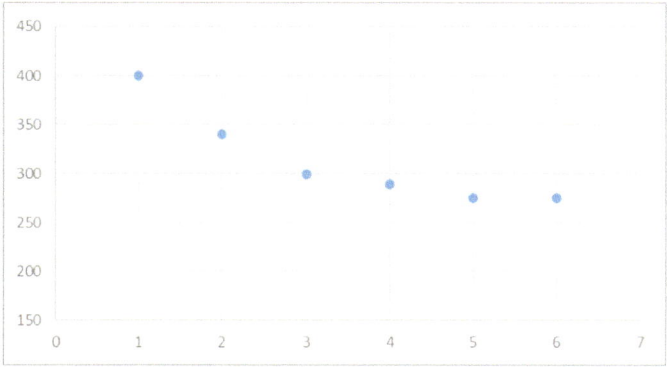

Fig. 15.7 Force values (kN) versus time in months

Fig. 15.8 Protection of the head of the tie rod. *Source* Rincent BTP—France

Chapter 16
Sampling

The representativeness of a sample is always only partially verifiable. It is a relative notion. A sample can be representative of one, two, three or more variables, but is never completely identical to the total population. Even if representativeness is fully verifiable, it is verified at a given time.

We chose a wall of 99 tie rods, which we will call 100 tie rods to reduce the analyses to percentages. This wall where all the tie rods were tested is representative of what we found during the tests, whatever the type of tie rod, bar or cables. Families of tie rodsFirst of all, there are families of tie rods attached to the lines, i.e. to their altimetric position, in this example there are 5 lines of tie rods.

Family 1.

See Figure 16.1.

Family 2.

Line 1.

See Figure 16.2.

Family 3.

Lines 3 et 4.

See Figure 16.3.

Family 4.

Line 2.

See Figure 16.4.

This line of tie rods has lost most of its initial tension. This observation is not a particular case, but a general case of tie rods involved in the support of walls that receive rolling loads.

This observation concerning line 2 of our example is sometimes extended to several lines of tie rods. This is the case, for example, for a 16.5 m high wall with 11 tie lines (Fig. 16.5).

Lines 3 and 4 have significantly lower force values than lines 2 and 5 (Fig. 16.6).

Line 4 has significantly lower tension force values than lines 3 and 5.

© The Author(s) 2024
J.-J. H. Rincent, *Ground Anchors*, https://doi.org/10.1007/978-981-97-4414-5_16

Family 1

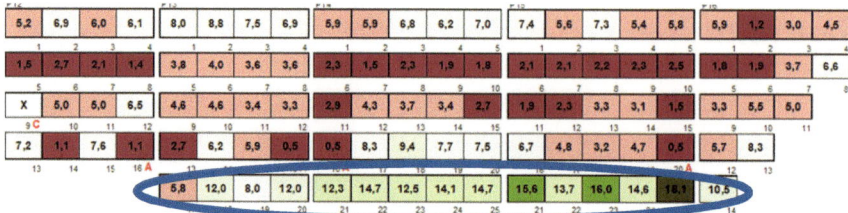

Fig. 16.1 Family 1

Family 2

Line 1

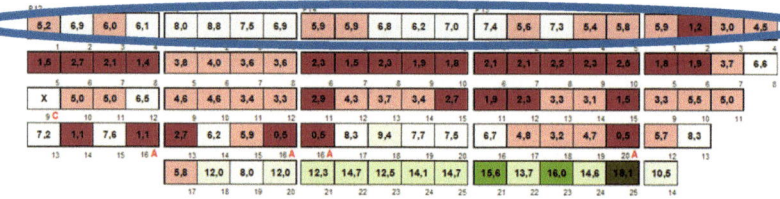

Fig. 16.2 Family 2

Family 3

Lines 3 et 4

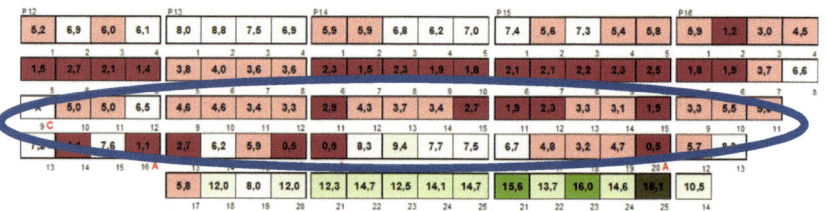

Fig. 16.3 Family 3

Laurent Soyez's thesis is: Contribution to the study of the behavior of retaining structures subjected to railway operating loads (dynamic and cyclic loads) 2009 (Figs. 16.7 and 16.8).

Numerous tests have been carried out on a reinforced embankment at a scale of 1:1. In practice, the following curves show that the load applied to the embankment is maximum between 1- and 2-m depth and close to zero at 3.5 m depth. The parameter concerning the duration of load application is directly related to the depth of load application (Fig. 16.9).

Family 4

Line 2

Fig. 16.4 Family 4

Fig. 16.5 Tension forces of the tie rods

7,2	0,7	3,8	1,6	0
10,4	10,1	8,6	11,3	0,9
3,5	8,7	1,4	10,8	2,25
4,6	0	1,1	11,5	3,63
10,9	11,3	10,2	14	5,56
13,9	12,9	12,2	12,2	7,38
15,9	13	14	14,9	9,31
18,7	23,2	23,6	18,4	10,88
23,6	13,4	20,5	23	12,5
6,4	24,5	18	24,4	14,13
13,1	16,9	23	34,2	16,5

Fig. 16.6 Force per line of tie rods

8	7,5	11,6	7	0
11,6	6,5	3,9	9,3	0,9
14,6	6,6	13,4	6,8	2,25
1,3	0	5,9	0,3	3,63
14,2	1,6	11,4	6,9	5,56
16,1	19,4	19,2	9,8	7,38
23,3	5,4	24,2	12,4	9,31
13,1	17	20,7	3,7	10,88
13,5	17	18,7	17,3	12,5
16,3	2,8	25,4	20,3	14,13
22,9	20,4	20,3	22,8	16,5
22,7	20,2	16,1	19,4	17,5

We carried out a static tensile test on a tie bar receiving cyclic loads due to a convoy of ore wagons and measured deformations in the tie bar, which are close to those shown in the previous figure for a load of 250 kN.

Sampling should also consider the following.

1. Boundary zones of the retaining wall where tensile forces are lowest, in blue.
2. These boundary zones of the wall are locations where in-place soils are found and fill soils decrease, resulting in lower tension values in the tie rods than in tie rods of the same line or level.

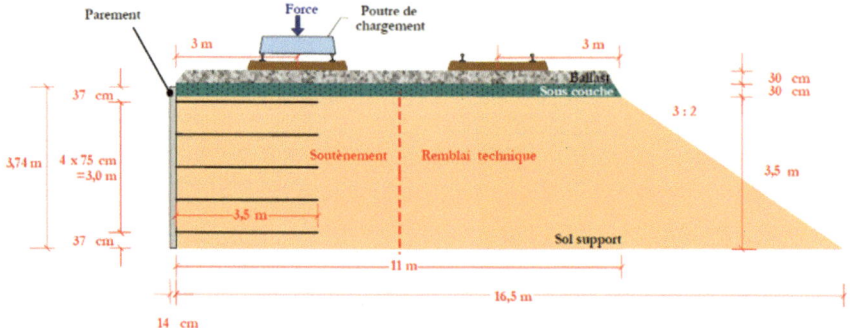

Fig. 16.7 Cross section of the test embankment. *Source* Laurent Soyez

Fig. 16.8 Diffusion of railway loads. *Source* Laurent Soyez

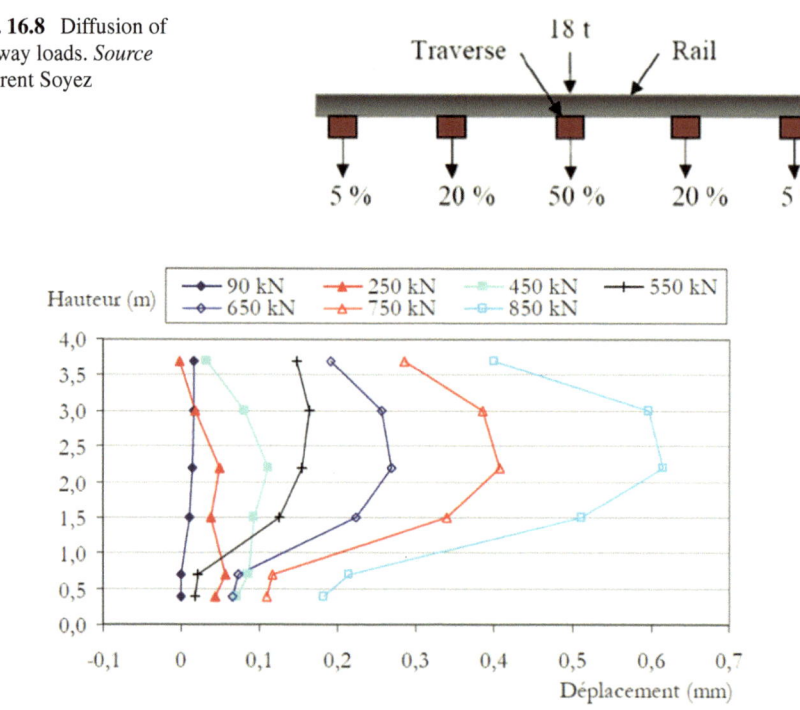

Fig. 16.9 Deformed as a function of load. *Source* Thesis de Laurent Soyez

3. Tie bars near construction joints blue and yellow arrows.

Construction joints between concrete wall panels have specific locations where rainwater or groundwater can flow, but in general, this flow or even seepage results in soil particles that create locally decompressed areas, hence the low-tension forces measured.

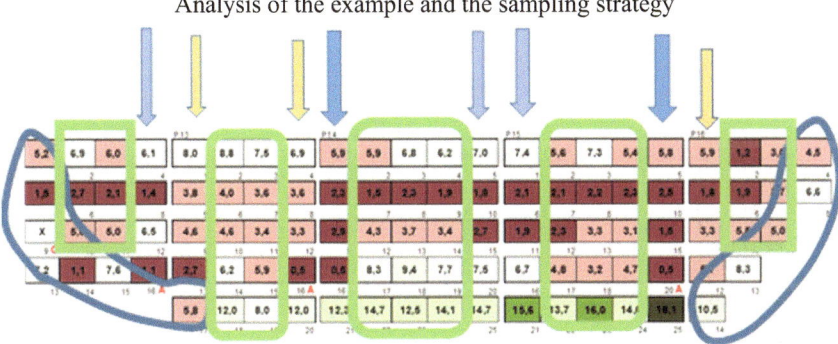

Analysis of the example and the sampling strategy

Fig. 16.10 Examples of specific points

4. Zones of "normal" behavior, green frames

The central parts of the wall, avoiding the proximity of particular points, give a correct representation of the behavior of the wall, in general, two types of tie rods, tie rods that support only static loads and tie rods that receive, in addition, dynamic cyclic loads from heavy transport vehicles.

Analysis of the example and the sampling strategy (Fig. 16.10).

For this wall used as a working example:

- 52% of the tie rods are in "normal" situations
- 11% are in the outer limit zones of the wall
- 37% are close to the construction joints

The sampling strategy is an essential step in the design of scientific experiments, with or without particular experimental treatment, i.e. including measurements on an object.

Cochran formula.

16.1 $N = t^2xp(1-p)/m^2$

N: minimum sample size to obtain significant results for a given event and risk level.

t: confidence level (the standard value for the 95% confidence level is 1.96).

p: estimated proportion of the population with the characteristic.

m: margin of error (usually set at 5%).

For anchors called "normal".

With a 95% confidence level, see Student's law and the value of n equal to 1.96. (Fig. 122). The proportion of the population with these characteristics is 68%.

For a margin of error of 10%, the number of tie rods to be tested to reach these objectives is **43**.

For the same population with lower assumptions, for example.

Loi de Student

Valeurs de t ayant la probabilité P d'être dépassées en valeur absolue.

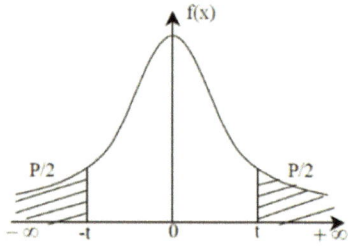

120	0,1259	0,2539	0,3862	0,5258	0,6765	0,8446	1,0409	1,2886	1,6576	1,9799	2,6174
200	0,1258	0,2537	0,3859	0,5252	0,6757	0,8434	1,0391	1,2858	1,6525	1,9719	2,6006
∞	0,1257	0,2533	0,3853	0,5244	0,6745	0,8416	1,0364	1,2816	1,6449	1,9600	2,5758

Fig. 16.11 Student's law

- A confidence level of 90% (n = 1.64)
- And a margin of error of 15%
- The minimum number of tests is **16**.

There are other more sophisticated methods of calculation, but this example shows the difficulty of choosing the items to be tested and the link between the number and its representativeness for a group (Fig. 16.11).

Chapter 17
Statics and Dynamics Tests

See Figure 17.1.

The objective of the static and dynamic tests carried out simultaneously is to solve the first-degree equations of the type $Y = Ax + B$ and thus to calculate A and B.

Y is the square root of the dynamic stiffness and x is the internal force of the tie rod (Fig. 17.2).

The behavior of the tie rod under tension is generally divided into three parts:

- Initial installation phase, interference from the weight of the equipment required for testing
- Phase in which the forces are lower than the internal tensile force of the tie rod, constant stiffness
- Phase in which the applied force is greater than this internal force

The type of curve shown is quite common. It should be noted that exceeding the internal tensile force (blue arrow) induces a change in the equilibrium of the system. The tie rod and the ground are stressed for values that induce changes in behavior.

Tie rod 141 example.

This tie rod is located on another wall without a concrete structure nearby (Figs. 17.3 and 17.4).

The static test is performed simultaneous with dynamics tests. The crescent and decrescent values of force are:

- 1,2t – 4t – 8t – 12t -and 16t
- 12t – 8t- 4t and 1,2t (Fig. 17.5).

The detachment of the support plate from the tie-head was observed under a load of 7 tons.

© The Author(s) 2024
J.-J. H. Rincent, *Ground Anchors*, https://doi.org/10.1007/978-981-97-4414-5_17

Fig. 17.1 Diagram of the
traction device

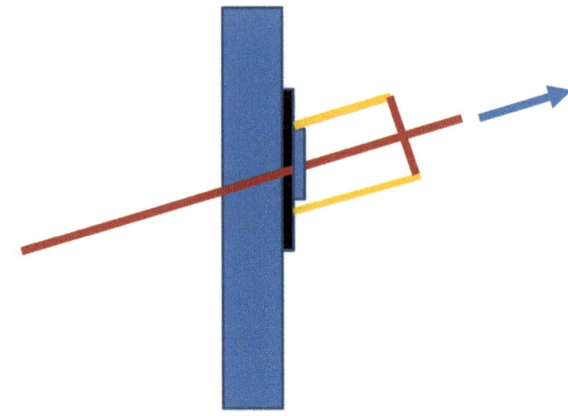

static and dynamic test simultaneous

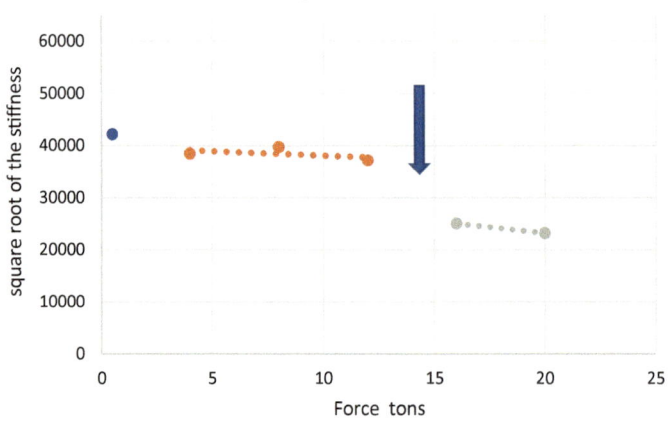

Fig. 17.2 Square root of the stiffness function of force

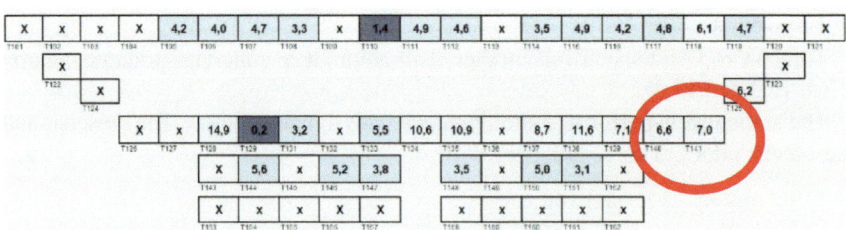

Fig. 17.3 Tie rod 141

Fig. 17.4 Tie rod 141

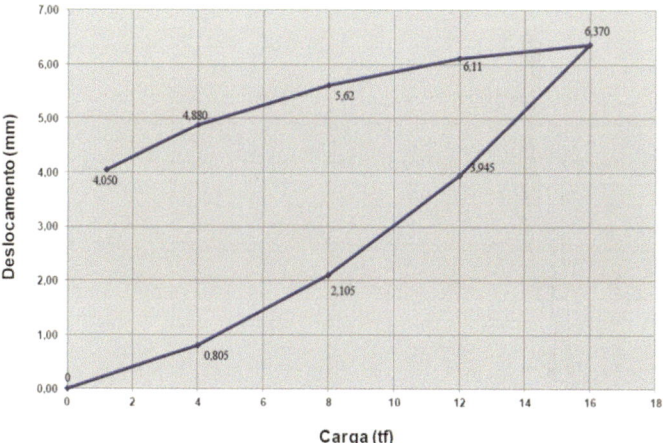

Fig. 17.5 Static test

Detailed dynamic stiffness results are given in Appendix 7 and are summarized in the following curve (Fig. 17.6).

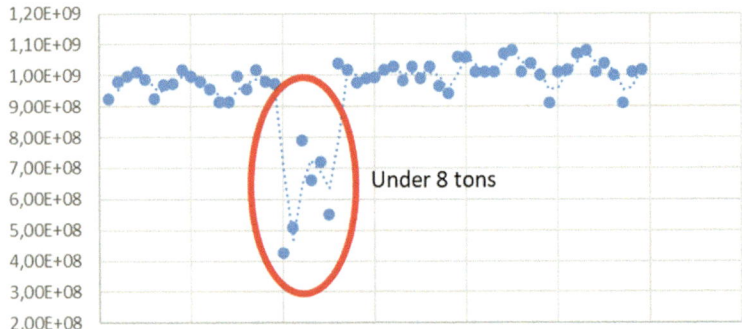

Fig. 17.6 Dynamic stiffnesses T141 under different loads

The points circled in red are characteristic of the new tie-rod load, since the tensile force exceeds the existing internal force.

Chapter 18
Accuracy of Measurement

The eight strikes made on the head of the tie rod reveal different behaviors of the tie rods.

In the first case, the tie bar can be stable under the stresses caused by the hammer.

For example, tie rod 115 which contributes to the stability of a wall, the 8 values of stiffnesses calculated.

7,31E+08 by removing the maximum and minimum stiffness

    7,19E+08 Average

    6,77E+08 6,80E+08 N/m

    6,70E+08

    6,66E+08 relative incertitude

    6,80E+08 1,21%

    6,35E+08

    6,66E+08

The tie rod is sensitive to the vibrations caused by the hammer.

The internal tension of the tie rod increases with the second stroke, decreases sharply with the third, and decreases again with the sixth stroke, and then decreases again with the 7th and 8th strokes.

- 2,30 E8 by removing the maximum and minimum stiffness
- 2,39 E8
- 8,23 E7 Average 1,53 E8 N/m
- 1,00 E8
- 1,65 E8 Relative incertitude 13,9%
- 1,95 E8
- 1,17 E8
- 1,13 E8

J.-J. H. Rincent, *Ground Anchors*, https://doi.org/10.1007/978-981-97-4414-5_18

Fig. 18.1 Force and incertitude in percent results. Average: 5,8%. Standard deviation: 0,64

F tons	% incertitude	F tons	% incertitude
11,6		7	
	9,8		2,6
3,9		9,3	
	3,3		4,6
13,4		6,8	
	7,6		1,88
5,9		0,3	
	3,04		6,6
11,4		6,9	
	12,8		11,2
19,2		9,8	
	1,76		10,4
24,2		12,4	
	4,9		11,3
20,7		3,7	
	5,7		25,7
18,7		17,3	
	1,36		7,7
25,4		20,3	
	1,24		8
20,3		22,8	
	8,9		7,1
16,1		19,4	
	2,6		4,8

We have chosen the example of a 12-level tensioned wall to carry out the calculation of incertitude on the calculated dynamic stiffnesses, we have chosen the acquisitions in two columns of tie rods.

For each tie rod, 8 acquisitions are made, the calculation of the 8 stiffnesses and will be kept 6 values, removing the smallest and largest value.

The calculation of the mean and the standard deviation was carried out for each of the 24 ties. The relative incertitude is calculated from the standard deviation of the mean value over the 6 values.

The relative incertitude in percent are as follows (Fig. 18.1):

The measurement which gives a relative incertitude of 25% is considered as a particular tie rod. The shocks caused by the hammer modify the behavior of the tie rod which has a low internal force which, in this case, is 3.7 tons.

This type of tie rod is unstable and fragile even under the impact of the hammer that generates the vibration wave. The 10 tie rods whose internal force is greater than 15 tons have a measurement relative incertitude of 5.1%.

Another example of a wall (Fig. 18.2).

10,9	11,3	11,5	10,4	10	9,3	9,6	10,6	11,3	13,2	11,8	10,5
0,7	4,1	4,6	1,1	2,7	2,1	1,3	1,64	2,47	4,47	1,27	0,58
19	20	21	22	23	24	43	44	45	46	47	48
13,9	13,1	12,9	12,7	11	10,5	11	11,5	10,9	10,1	10,9	9,7
1,3	0,5	0,6	1,8	1,4	0,9	1	0,68	1,78	1,96	2,11	1,21
25	26	27	28	29	30	49	50	51	52	53	54
14,4	15,3	17	13,4	13	7	12	11	9,75	10,3	9,9	8,7
0,6	0,5	0,9	3,4	0,5	8,7	0,7	2,24	1,35	1,76	2,15	3,02
31	32	33	34	35	36	55	56	57	58	59	60
19	21,9	2,4	17,3	12	13,5	15	13,2	10	10,8	9,5	9,6
1,5	2,9	9	8	6,4	1,7	1,6	8,33	5,57	1	3,11	0,74
37	38	39	40	41	42	61	62	63	64	65	66

Force in tons
Incertitude %
Tie rod number

Fig. 18.2 Two panels of a wall, effort and relative incertitude

Fig. 18.3 Photography of the same wall

84% of the relative incertitude are lower than 5%, when the values are higher than 5%, these results concern particular tie rods in situations and location, for example, construction joints, water outflow, etc. (Fig. 18.3).

Chapter 19
Examples

From 2019 to 2021, the Rincent BTP Services branches in Recife and Sao Paulo, Brazil, carried out more than 2,000 tests on tie-rods.

One of the two major projects involved 19 retaining walls representing a total of 1,811 tie rods. On the largest embankment, 590 out of 741 tie rods were tested using dynamic tests.

The static tests help to exploit the dynamic tests and provide information on the behavior of the tie-rods with a view to re-tensioning (Fig. 19.1).

In this study, more than 20 static tests were carried out with the following objectives:

– Determine the tension force in the tie rods to calibrate the dynamic tests
– Analyze their behavior under tensile stress
– Simultaneously analyze the dynamic stiffnesses measured at each load level.

Tie rod 102

The V/F curves as a function of frequency, on which static and dynamic tests were carried out. For 590 elements tested, 4720 curves of this type were analyzed (Fig. 19.2).

The curve for tie rod 102 gives:

– a stiffness of 1.28 E9 N/m,
– a total length of 23.75 m with a wave velocity in the tie rod of 4500 m/s.
– the free length is 14,4m for the same hypothesis.

The admittance of 1.1 E-6 m/sN leads to a calculation of the equivalent diameter of 0.33m, strands reinforcement plus grout.

Tie rods 138 – 261 – 3127 – 3132 8 strands
138 – 450 – 451 – 5125 – 5138 10 strands

The table of the appendix 5 shows the average stiffness values and the relative incertitude of these measurements, enabling us to draw up the following summary table. Appendix 6 shows the location of tie rods.

© The Author(s) 2024
J.-J. H. Rincent, *Ground Anchors*, https://doi.org/10.1007/978-981-97-4414-5_19

Fig. 19.1 Wall of 590 tie rods. *Source* Rincent BTP Recife

The physical significance of high relative incertitude values is that the tie-rod is sensitive to the compression wave generated by the hammer, or more simply that it is brittle. However, a completely relaxed tie-rod will give results with quite variable dynamic stiffnesses.

These tie rods are subjected to static tests, with dynamic tests carried out at each stage during loading and unloading.

The results of these simultaneous tests refine the calculation of a and b, and enable us to establish the formulas which, for the case presented, were:

- Y = 1750x + 7500 pour 8 strands
- Y = 1750x + 11000 pour 10 strands

See Figures 19.3 and 19.4.

Tie rod 709 Bis

The aim of the tests carried out on this tie was to reach the breaking force to validate the possibility of re-tensioning part of the tie.

The dynamic stiffnesses measured on this tie were homogeneous, leading to a calculation of the tension force of 9.38 tons using the formula for 8 strands (Fig. 19.5).

The static test showed that the distribution plate separated from the head of the tie-rod at a force of 9.6 tons. The tensile test was carried out up to 42,5 tons without failure.

Electrical insulation measurement tests were carried out in accordance with SIA 267/1 (2013), which states:

10.7.4.2 A tie rod, once injected and tensioned, must have an electrical resistance RI 0.1 MΩ (Mega Ohm) (Fig. 19.6).

The values of 6 MΩ are well above the required value (Fig. 19.7).

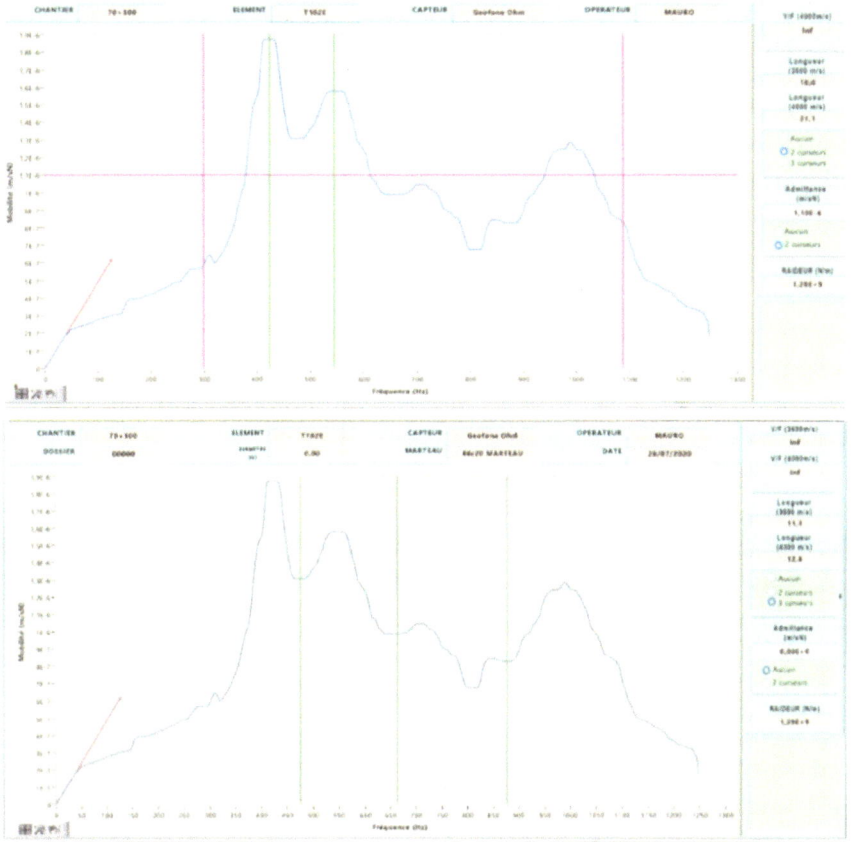

Fig. 19.2 V/F function of the frequency

It appears that the relative inc7rtitude of stiffness measurements of less than 10% and compliance with the electrical insulation criterion are two criteria that must be met before the tie rods can be re-tensioned.

Tie rods 450 and 451 (Fig. 19.8).

These two tie rods, whose internal tension is close to zero or null, have both been tensioned up to 45 tons (Fig. 19.9).

Tie rods 518 and 5134

The non-destructive and static tests carried out on tie rods 418 and 5134 confirm the preliminary approach to determining which tie rods can be re-tensioned (Fig. 19.10).

The static test showed that the distribution plate separated from the head of the tie-rod at a force of 9.6 tons. The tensile test was carried out up to 53 tons without failure.

Electrical insulation measurement tests were carried out in accordance with SIA 267/1 (2013), which states:

Fig. 19.3 Results table

8 strands	Dyn. Stiffness	Relative incertitude	
238	$2,10^E9$	4,24%	
261	$2,08^E8$	4,62%	Close to a construction join
3127	6,06E8	65%	
3132	$2,42^E9$	3,40%	
10 strands			
138	2,21E9	2,86%	
450	1,68E8	16,13%	Tension force close to zero but in contact with the wall
451	1,12E8	24,46%	Zero Tension Force\n\nTie-rod only
5125	3,08E9	4,74%	
5138	4,78E9	11,77%	Close to the water collection structure

Fig. 19.4 Example of a concrete structure near the tie rod, foot of the wall

Fig. 19.5 Dynamic
stiffnesses N/m table

Rd	
5,41E+08	
5,66E+08	
5,64E+08	
5,79E+08	
5,88E+08	
5,91E+08	5,72E+08
	3,25%

TIRANTE N° T709 Bis	FIO 1	FIO 2	FIO 3	FIO 4	FIO 5	FIO 6	FIO 7	FIO 8
LEITURA 1 (MΩ)	7	7	6	6	6	6	6	6
LEITURA 2 (MΩ)	6	7	6	6	6	6	6	6
LEITURA 3 (MΩ)	6	6	6	6	6	6	6	6

Fig. 19.6 Electrical isolation measurement

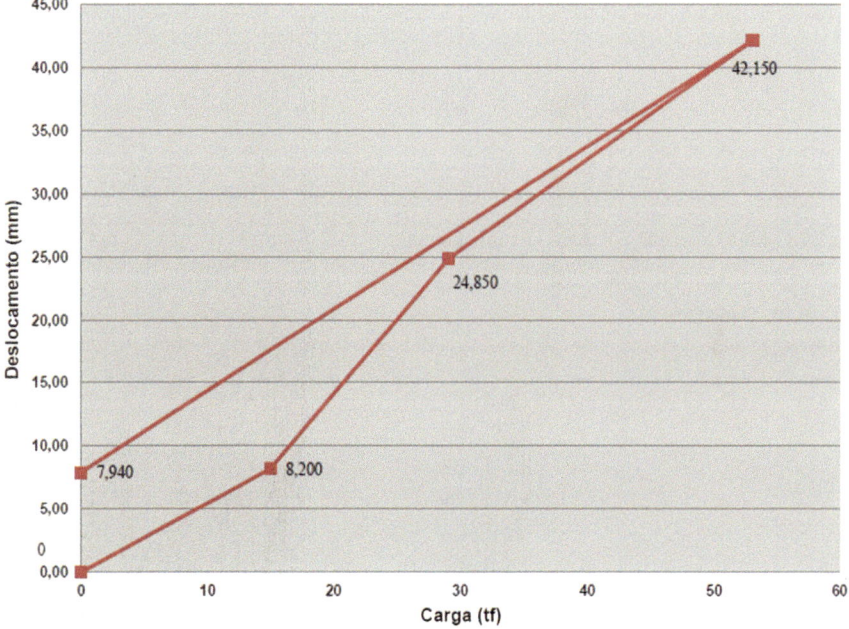

Fig. 19.7 Static test 709bis

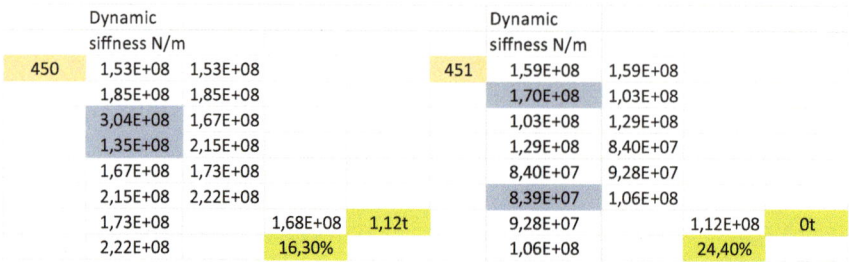

Dynamic siffness N/m				
450 1,53E+08	1,53E+08			
1,85E+08	1,85E+08			
3,04E+08	1,67E+08			
1,35E+08	2,15E+08			
1,67E+08	1,73E+08			
2,15E+08	2,22E+08			
1,73E+08		1,68E+08	1,12t	
2,22E+08		16,30%		

Dynamic siffness N/m				
451 1,59E+08	1,59E+08			
1,70E+08	1,03E+08			
1,03E+08	1,29E+08			
1,29E+08	8,40E+07			
8,40E+07	9,28E+07			
8,39E+07	1,06E+08			
9,28E+07		1,12E+08	0t	
1,06E+08		24,40%		

Fig. 19.8 Results table

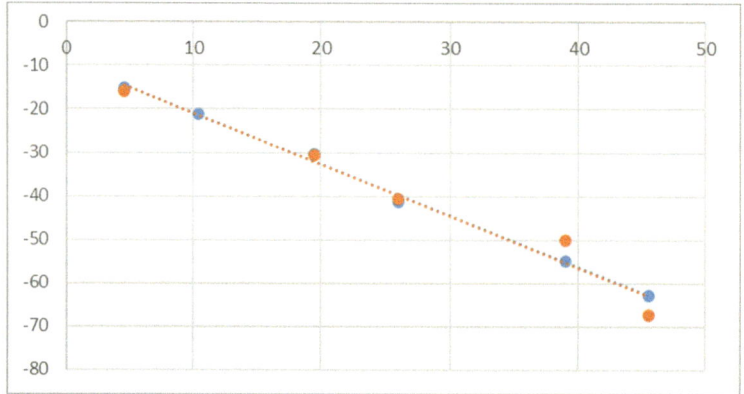

Fig. 19.9 Re-tension curve F tons—deformation mm

Fig. 19.10 Dynamic stiffnesses table

418 Rd		5134 Rd	
7,42E+08		1,32E+09	
6,59E+08		1,46E+09	
8,64E+08		1,41E+09	
1,12E+09		1,34E+09	
5,45E+08		1,43E+09	
7,54E+08		1,17E+09	
	7,81E+08		1,36E+09
	25,22%		7,72%

10.7.4.2 A tie rod, once injected and tensioned, must have an electrical resistance RI 0.1 MΩ (Mega Ohm).

The values of 6 MΩ are well above the required value.

It appears that the relative incertitude of stiffness measurements of less than 10% and compliance with the electrical insulation criterion are two criteria that must be met before the tie rods can be re-tensioned.

The non-destructive and static tests carried out on tie rods 418 and 5134 confirm the preliminary approach to determining which tie rods can be re-tensioned.

The calculated forces in these ties are 9.7t for tie 418 and 13.1t for 5134. The major difference between the two tie rods is that tie rod 418 is sensitive to non-destructive testing, since the relative incertitude is well over 10% (Figs. 19.11 and 19.12).

Dynamic tests show the new functioning of the tie-rod, which is loaded at a force value higher than its internal tension force.

T 308

This tie-rod, located at the foot of a motorway retaining wall, was subjected to a static tensile test and simultaneous dynamic tests at each loading level.

Initial analysis of the test curve revealed:

– a permanent deformation of 13.09 mm
– an elastic deformation of 4.8 mm measured on unloading after having reached 31 tons in tension (Fig. 19.13).

In practice, this means that the 31tons load resulted in permanent deformation of the grout cylinder and deformation under load, corresponding to a much lower load than normal on a ground anchor of this type.

The unloading curve is virtually linear, corresponding to the elastic behavior of the loaded part of the ground anchor armature.

A calculation of the tie-rod length mobilized from the elastic deformation yields a length of 4.3 m.

Dynamic tests recorded under a 31-tons load show a vibratory response of 4.4 m for an assumed wave velocity of 4,000 m/s (Fig. 19.14).

The total length of the tie rod is 15 m

This example was chosen to show that static testing can produce erroneous results, such as the internal tension of the tie-rod being greater than 31 tons, whereas the test only stressed a part of the tie-rod located at the rear of the wall. Further analysis of the dynamic tests reveals the presence of a grout bulb immediately behind the retaining wall.

Tie rods 412 – 504 – 567 – 580.

These 8-strands tie rods were re-tensioned after electrical isolation measurements in compliance with SIA 267/1 (Figs. 19.15 and 19.16).

The first re-tensioning operations were successfully carried out with limited effort. A key challenge is to design a tensioning device that can be attached to the head of the tie rod. The fact of working at height must also be taken into account.

The test campaign presented here enables us to map the internal force of the tie rods at a given point in time, based on a limited number of static tests and dynamic tests on a large number of tie rods. All these tests, together with electrical isolation measurements, enable us to select the tie rods that can be re-tensioned. This type of investigation leads to a pre-costing of the work to be carried out, optimizing it.

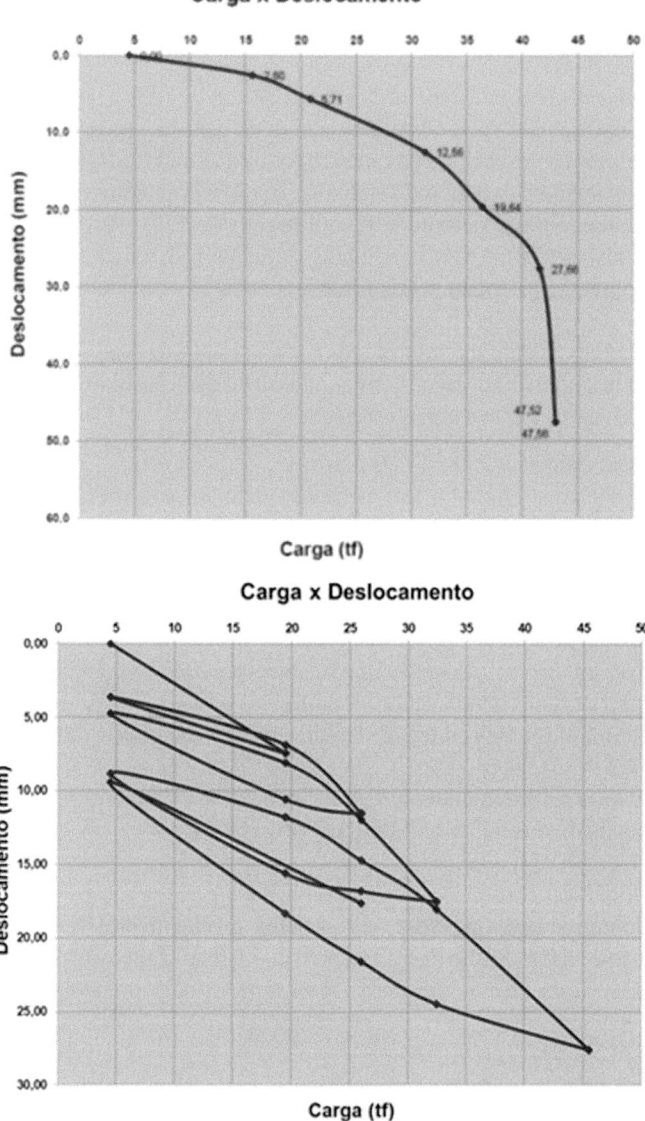

Fig. 19.11 Static tests

Fig. 19.12 Dynamic stiffnesses under different loads

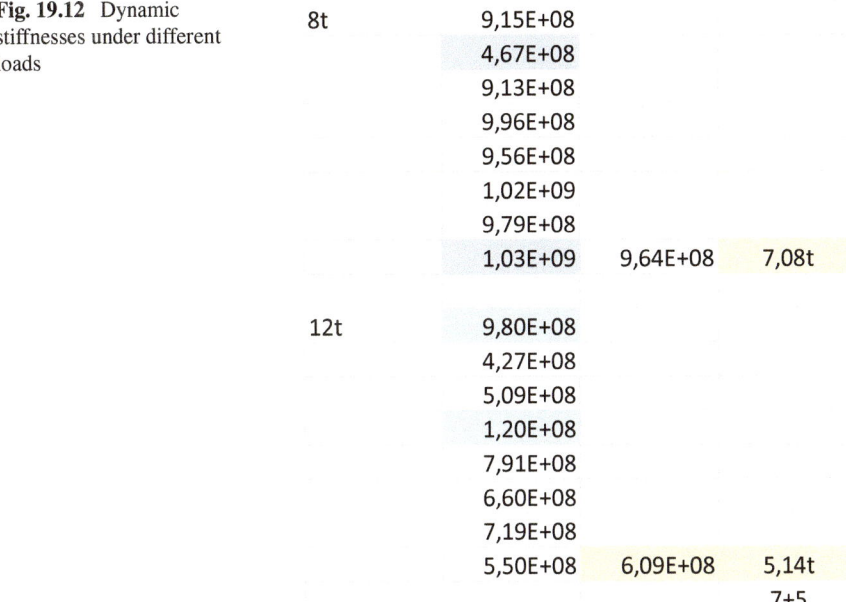

8t	9,15E+08		
	4,67E+08		
	9,13E+08		
	9,96E+08		
	9,56E+08		
	1,02E+09		
	9,79E+08		
	1,03E+09	9,64E+08	7,08t
12t	9,80E+08		
	4,27E+08		
	5,09E+08		
	1,20E+08		
	7,91E+08		
	6,60E+08		
	7,19E+08		
	5,50E+08	6,09E+08	5,14t
			7+5

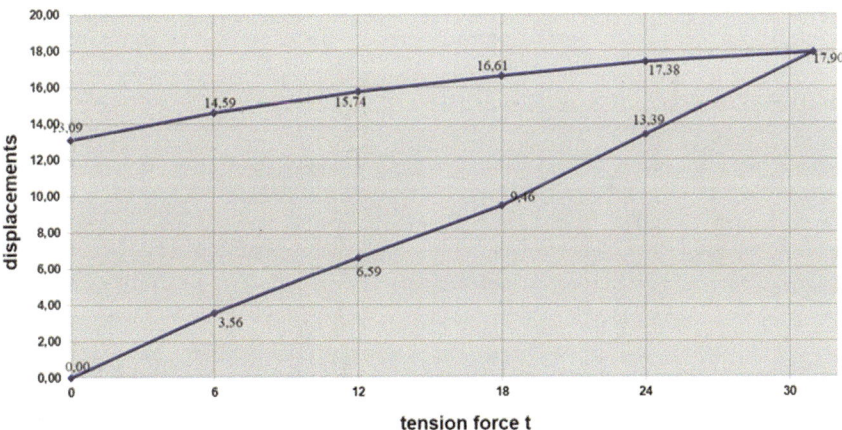

Fig. 19.13 Load and unload of a static test

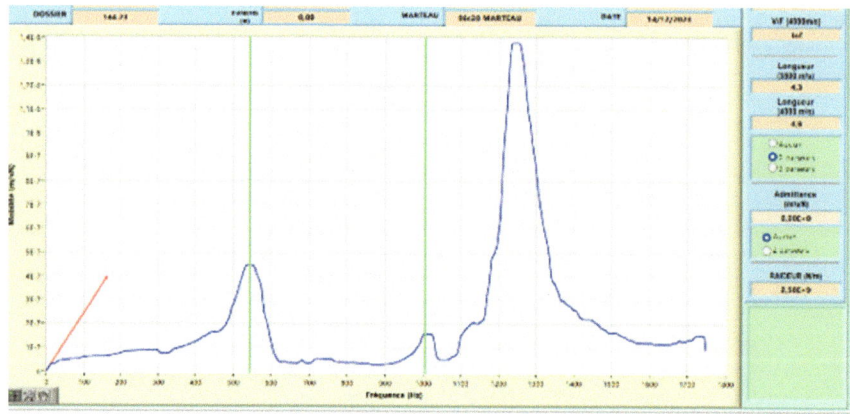

Fig. 19.14 Response

Fig. 19.15 Results table

N°	Tension force in tons	Dynmic stiffness relative incertitude
412	7,5	11,03%
504	7,2	8,91%
567	25,3	25,60%
580	12,4	27,94%

Fig. 19.16 Re-tensioning operation

Chapter 20
Conclusions

Based on the experience gained over 20 years of testing, there are a number of essential points to be borne in mind when mastering this method.

The first concerns the fixing of the geophone on the element to be tested. This fixing must enable the dynamic stiffness of the tested element to be measured, not the detachment of the geophone under repeated strikes on the element.

The compression waves generated by the hammer during the 8 acquisitions improve the relative uncertainty of the 6 results selected, and also enable us to assess the behavior of the tie-rod, its sensitivity to shocks and, by the same token, the possibility of re-tensioning it.

The other important point is sampling. For example, on a site where the tie rods have 8 or 10 strands, it is necessary to establish two formulas linking dynamic stiffness and force, as the inertias of the tie rods are different. The same applies to the thickness of the retaining wall, shotcrete and any other elements that may influence the relationship between dynamic stiffness and tension force.

As far as corrosion is concerned, the Swiss standard used for new tie rods can also be used for old tie rods.

Last but not least, re-tensioning tie rods is an important factor in optimizing maintenance operations. To facilitate testing and maintenance of tie rods, it would be advisable to upgrade the tie-head protection system.

Finally, all the results show the influence of fatigue phenomenon on load losses and tie-rod embrittlement.

The non-destructive method presented here has been in operation for over 20 years. It can be used to diagnose the condition of tie rods used to reinforce structures.

These tests enable us to.

- Determine whether the tie-rod is broken or not
- Calculate total tie rod length
- the free length
- the internal tension force at the time of testing.

© The Author(s) 2024
J.-J. H. Rincent, *Ground Anchors*, https://doi.org/10.1007/978-981-97-4414-5_20

By analyzing the curves, we can check whether the tie rods conform to the construction drawings, when they exist, and also calculate the diameters of the tie rods with the grout bonded to the ground, which leads us to check the dimensioning of the tie rods.

A limited number of static tests calibrate these non-destructive tests, and electrical isolation measurements complete the diagnosis.

All these tests lead to an evaluation of the existing situation, and to proposals for maintenance operations that are both technically and financially optimized.

This diagnosis makes it possible to identify the tie rods where a re-tensioning is possible and those too fragile which will not support this type of operation.

This technical contribution leads to the optimization of the work to be done.

The diagnostics contribute to the maintenance of the retaining walls, allow to follow the evolution of the tie rods and to plan the re-tensioning.

The ultimate goal is to guarantee the stability of the structures over time.

Appendix A

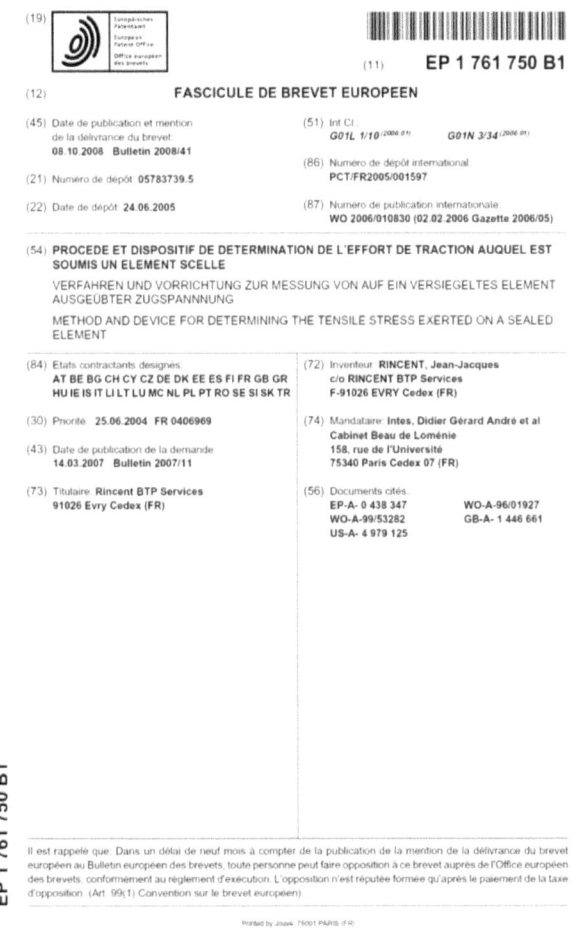

EP 1 761 750 B1

Description

[0001] L'invention concerne un procédé et un dispositif de détermination de l'effort de traction auquel est soumis un élément scellé maintenant une structure contre un support, l'élément scellé étant contraint en traction.

[0002] Les éléments scellés permettent de reprendre des efforts de traction dans le domaine du bâtiment, des travaux publics et de l'industrie. Ces éléments scellés sont en particulier des tirants actifs (précontraints) ou passifs qui participent à la stabilité des ouvrages du type murs de soutènement ou de quais en site portuaire, des boulons utilisés pour assurer la stabilité des voûtes des tunnels, des clous mis en oeuvre dans les renforcements de sols dans les talus ou des chevilles précontraintes fixant des éléments contre un support.

[0003] Il est connu de déterminer les efforts dans les éléments scellés en réalisant des essais destructifs, en particulier des essais de mise en traction directe.

[0004] Cependant ces essais de mise en traction directe impliquent un aménagement et une préparation de l'élément à tester en ré-usinant in situ la partie accessible, puis en fixant le dispositif d'essais par une connexion pouvant être par exemple un filetage. Il est ensuite nécessaire de mettre l'élément en tension pour déterminer l'effort dans ce dernier. Or, cette mise en tension peut être dangereuse, notamment lorsque l'effort atteint celui de la rupture. En effet les éléments scellés étant généralement constitués de métal, ils se corrodent et se fragilisent au cours du temps, en particulier lorsqu'ils se trouvent en milieu agressif, de sorte que les essais peuvent conduire à la rupture et mettre ainsi les opérateurs d'essais en danger ; en outre, la rupture de l'élément scellé peut entraîner des désordres sur l'ouvrage testé.

[0005] Des procédés et dispositifs selon l'état de la technique sont décrit dans les documents US 4 979 125, EP 0 438 347, WO 99/53282 et GB 1 446 661.

[0006] Le but de l'invention est de fournir un procédé et un dispositif qui permettent de déterminer l'effort de traction auquel est soumis l'élément scellé à partir d'essais non destructifs. Les avantages que procure la détermination de l'effort à partir d'essais non destructifs sont l'intérêt économique de ce type de procédé, la réalisation de ces essais non destructifs sur des sites où des essais destructifs sont difficilement réalisables, entre autre sous l'eau, et le fait que ce type d'essais est sans danger.

[0007] Le but est atteint par le fait que le procédé de détermination de l'effort de traction auquel est soumis un élément scellé comporte les étapes suivantes :

a) on fournit une loi de variation entre la raideur dynamique de l'élément scellé et un effort statique de traction auquel ce dernier est soumis,

b) on soumet l'élément scellé à un impact de force déterminée, pour générer une vibration de l'élément scellé et de la structure,

c) on relève la réponse vibratoire de l'élément scellé et de la structure mis en vibration,

d) on détermine la raideur dynamique de l'élément scellé et de la structure à partir de la réponse vibratoire,

e) on corrige la raideur dynamique de l'élément scellé et de la structure d'un paramètre traduisant l'influence de la structure sur cette dernière raideur, et

f) on applique la loi de variation à cette raideur dynamique corrigée pour obtenir l'effort dans l'élément scellé.

[0008] Ainsi, le procédé permet de retrouver l'effort de traction auquel est soumis un élément scellé, à partir d'essais non destructifs. En l'espèce, l'effort de traction est déterminé à partir d'une loi de correspondance entre effort statique et raideur dynamique et à partir d'un essai dynamique (ou plus précisément à partir de l'analyse de la réponse vibratoire de l'élément mis en vibration suite à un impact de force déterminée). L'impact de force est déterminé, par exemple en mesurant l'amplitude de cette force au moment de l'impact.

[0009] Avantageusement, cet impact peut être réalisé sur la tête de l'élément scellé ou, en particulier lorsque l'élément scellé comporte une plaque d'appui.

[0010] Avantageusement, après avoir effectué un traitement mathématique de la réponse vibratoire de l'élément scellé et de la structure mis en vibration, on construit une courbe qui correspond au rapport de la vitesse de propagation sur la force appliquée en fonction de la fréquence et on détermine la raideur dynamique de l'élément scellé et de la structure en mesurant sur cette courbe l'inverse de la pente à l'origine multiplié par 2π.

[0011] L'invention concerne aussi un dispositif de détermination de l'effort de traction auquel est soumis un élément scellé qui comporte :

- une mémoire contenant une loi de variation entre la raideur dynamique de l'élément scellé et un effort statique de traction auquel ce dernier est soumis,
- des moyens pour soumettre l'élément scellé à un impact de force déterminée pour générer une vibration de l'élément scellé et de la structure,
- des moyens pour relever la réponse vibratoire de l'élément scellé et de la structure mis en vibration,
- des moyens pour déterminer la raideur dynamique de l'élément scellé et de la structure à partir de la réponse

2

EP 1 761 750 B1

vibratoire,

- des moyens pour déterminer la raideur dynamique de l'élément scellé permettant de corriger la raideur dynamique de l'élément scellé et de la structure d'un paramètre traduisant l'influence de la structure sur cette dernière raideur, et
- des moyens pour appliquer la loi de variation à cette raideur dynamiques corrigée pour obtenir l'effort dans l'élément scellé.

[0012] Avantageusement, les moyens pour soumettre l'élément scellé à un impact de force déterminée comportent un marteau d'impact équipé d'un capteur de force et les moyens pour relever la réponse vibratoire comportent un capteur de vitesse, du type géophone ou vélocimètre, ou un accéléromètre.

[0013] Avantageusement, les moyens pour relever la réponse vibratoire sont positionnés sur l'élément scellé ou éventuellement sur la plaque d'appui de l'élément scellé, selon les conditions du site.

[0014] Avantageusement, le dispositif comporte des moyens de traitement mathématiques destinés à traiter la réponse vibratoire de l'élément scellé et de la structure.

[0015] Avantageusement, les moyens de traitement mathématiques comportent des transformées de Fourier, qui permettent de traiter la réponse vibratoire pour obtenir une courbe représentant la vitesse sur la force, en fonction de la fréquence.

[0016] L'invention sera bien comprise et ses avantages apparaîtront mieux à la lecture de la description détaillée qui suit, de modes de réalisation de l'invention représentés à titre d'exemples non limitatifs.

[0017] La description se réfère aux dessins annexés sur lesquels :

- la figure 1A représente schématiquement la tête d'un élément scellé maintenant une structure,
- la figure 1B représente en coupe l'élément scellé de la figure 1A,
- la figure 2 représente la tête d'un autre élément scellé,
- la figure 3 représente schématiquement un essai dynamique,
- la figure 4 représente une courbe illustrant la réponse vibratoire d'un élément mis en vibration,
- la figure 5 représente la loi de variation entre raideur dynamique et effort statique,
- la figure 6 représente la détermination de l'effort selon un premier mode de réalisation, et
- la figure 7 représente la détermination de l'effort selon un autre mode de réalisation.

[0018] La figure 1A illustre un élément scellé 10, en l'espèce un tirant précontraint 12 qui participe à la stabilité d'une structure composée d'une pluralité d'éléments de structure, en plaquant un élément de structure 14 contre un support. Pour toute la suite, l'élément de structure est désigné par structure. Le tirant 12 illustré sur la figure 1A, comporte une plaque d'appui 16 et une tête d'ancrage 18. En fait, l'élément dont on cherche à déterminer l'effort en traction pourrait être un tirant à torons 10', comme illustré sur la figure 2 comportant une plaque d'appui 16' et une tête d'ancrage 18', ou tout autre type d'éléments scellés connus, comme en particulier des boulons, des clous, des chevilles précontraintes, etc.

[0019] Le tirant 12, dont on aperçoit l'extrémité libre 12A sur la figure 1A est mis en place dans l'ouvrage de la manière suivante.

[0020] En référence à la figure 1B, après avoir effectué un forage 20 dans un support 22 qu'il convient de stabiliser, on introduit l'armature 12B du tirant 12 jusqu'à ce que l'extrémité 12C opposée à l'extrémité libre 12A se trouve dans un coulis de scellement 24. La plaque d'appui 16, qui est éventuellement disposée entre la tête d'ancrage 18 et l'élément de structure 24 destiné à stabiliser le support 22, et le serrage de la tête d'ancrage 18, permettent de précontraindre le tirant 12 sous une force connue.

[0021] Le dispositif selon l'invention comporte des moyens pour soumettre l'élément scellé à un impact de force Fd comme illustré sur la figure 3, l'impact pouvant être réalisé soit sur la plaque d'appui 16, soit directement sur la tête de l'élément scellé correspondant à l'extrémité libre 12A du tirant 12. En l'espèce, ces moyens soumettant l'élément scellé à un impact, comportent un marteau d'impact 26 équipé d'un capteur de force permettant de mesurer la force Fd appliquée, comme illustré sur la figure 1A. Le dispositif comporte en outre des moyens pour relever la réponse vibratoire de l'élément scellé 10 et de la structure 14, tous deux mis en vibration, qui comprennent un capteur de vitesse 28 du type géophone ou vélocimètre, ou un accéléromètre. Le dispositif comporte par ailleurs, des moyens d'acquisition et de traitement 30 permettant de réaliser des transformées de Fourier T de la réponse vibratoire 32 de l'élément scellé 10 et de la structure 14 et ainsi traiter cette réponse afin de pouvoir l'analyser.

[0022] Nous allons tout d'abord exposer la manière dont un essai dynamique est conduit et l'analyse qui en est faite.

Etape b)

[0023] On soumet la tête 12A du tirant 12 ou la plaque d'appui 16 à un impact de force Fd à l'aide du marteau d'impact 26, pour générer une vibration de l'élément scellé 10 et de la structure 14.

3

EP 1 761 750 B1

Etape c)

[0024]   Après avoir réalisé cet essai dynamique, on enregistre la réponse vibratoire 32 (figure 1A) correspondant à la réponse vibratoire de l'élément scellé 10 et de la structure 14 mis en vibration à l'aide d'un capteur de vitesse 28 et des moyens d'acquisition et de traitement 30.

[0025]   La réponse vibratoire 32 correspond en fait à une courbe 32A qui représente le signal de force d'impact Fd en Newton (N), en fonction du temps t en secondes (s), et d'une courbe 32B qui représente le signal de vitesse V de propagation de la vibration générée, en mètre/seconde (m/s), en fonction du temps t en secondes (s).

étape d)

[0026]   Ces deux courbes de réponse 32A et 32B sont ensuite traitées, à l'aide d'un traitement mathématique connu implanté dans les moyens d'acquisition et de traitement 30, comportant en particulier un traitement par une transformée de Fourier et les moyens d'acquisition et de traitement 30 permettent alors de tracer une courbe traitée 34, représentant la vitesse V de propagation rapportée sur la force Fd appliquée, V/Fd (en m/sN), en fonction de la fréquence f en Hertz (Hz), comme illustré sur la figure 4.

[0027]   La raideur dynamique R de l'élément scellé 10 et de la structure 14, est déterminée à partir de la courbe traitée 34 en déterminant l'inverse de la pente à l'origine (multiplié par 2Π). Le dispositif comporte à cet effet des moyens 36 pour déterminer cette raideur dynamique R à partir de la courbe traitée 34. Ainsi, l'allure de la courbe 34 au voisinage de l'origine est proche d'un segment rectiligne, qui s'étend entre l'origine et un point de coordonnées ayant pour abscisse et ordonnée, respectivement (β, α).

[0028]   Dans ce cas,

$$R = \frac{\beta}{\alpha} \times 2\pi .$$

Etape a)

[0029]   Pour pouvoir déterminer l'effort de traction auquel est soumis l'élément scellé 10, il convient alors de fournir une loi de variation L entre la raideur dynamique R de l'élément testé et un effort statique de traction F auquel il est soumis. Cette étape est une étape préliminaire réalisée à partir d'essais effectués avant les mesures selon l'invention.

[0030]   On a constaté que la correspondance entre la raideur dynamique et l'effort statique peut être approchée par une loi L de proportionnalité entre la racine carrée de la raideur dynamique et l'amplitude de l'effort. Cette loi L est :

$$(Rd)^{\frac{1}{2}} = \underline{a} \times F + (Ri)^{\frac{1}{2}}$$

où Rd représente la raideur dynamique de l'élément testé,
Ri représente la raideur de l'élément testé,
F représente l'effort dans l'élément testé, et
$\underline{a}$ dépend de l'élément testé.

[0031]   Notons que lorsque l'élément testé, pour déterminer cette loi L, est un élément scellé seul, Ri représente la raideur dynamique de l'élément scellé sous effort nul.

[0032]   Ainsi, cette loi L a été déterminée en testant un certain nombre d'éléments scellés 10 seuls analogues à ceux qui font l'objet des mesures selon l'invention, mais ne soutenant pas d'élément de structure contre un support, à l'aide d'essais statiques de traction et d'essais dynamiques.

[0033]   Une base de données a été créée en effectuant successivement des essais dynamiques sur un élément scellé 10 seul. En l'espèce, pour chaque essai dynamique, l'élément scellé 10 est soumis à une force d'impact Fd, tout en étant maintenu sous contrainte sous une force de traction donnée F, dont la valeur est distincte pour les différents essais.

[0034]   Pour chaque force de traction F appliquée, l'essai dynamique permet de déterminer, comme indiqué dans les étapes b) à d), la raideur dynamique R de l'élément scellé 10 correspondant, de manière à obtenir une base de données de couples de points effort statique/raideur dynamique (F, R).

[0035]   Après analyse de cette base de données, il a été observé qu'en reportant sur un graphe G1, comme représenté sur la figure 5, les valeurs de la racine carrée des raideurs dynamiques déterminées, en fonction des efforts statiques appliqués F en Newton (N), on obtient la loi L de proportionnalité telle que précitée.

4

EP 1 761 750 B1

[0036]   La valeur à l'origine de cette loi, représente la raideur dynamique intrinsèque R0 de l'élément scellé 10 seul (sous effort de traction nul). La valeur de cette raideur peut être obtenue en effectuant comme exposé précédemment un essai dynamique conduit selon les étapes b) à d) précitées. La raideur dynamique intrinsèque R0 de l'élément scellé 10 est déterminée sous effort nul, c'est-à-dire avant que l'élément scellé soit contraint en traction.

[0037]   La valeur de la raideur dynamique intrinsèque R0 de l'élément scellé 10 est alors reportée sur le graphe G1, sur l'axe des ordonnées, puisqu'elle représente la raideur dynamique sous effort nul, c'est-à-dire la valeur à l'origine.

[0038]   La pente a de la droite est déterminée en traçant une droite D (R0, a) ayant pour origine R0 et passant au plus proche des points relevés lors des efforts statiques de traction cumulés aux essais dynamiques sur les éléments scellés seuls 10.

[0039]   Lorsqu'il n'est pas possible de déterminer la valeur de la raideur dynamique intrinsèque R0 de l'élément scellé 10 seul à l'aide d'essais, il convient de tracer la droite qui passe au plus près des couples de points du graphe G1. Après avoir tracé une telle droite comme illustré sur la figure 5, il suffit de relever sa pente et sa valeur à l'origine, puisque cette dernière correspond à la raideur dynamique intrinsèque R0 de l'élément scellé 10 seul pour définir l'équation de la droite D (R0, a).

[0040]   Cette loi L ainsi complètement définie est implantée dans une mémoire M du dispositif.

Etape e)

[0041]   Cependant, la raideur dynamique R de l'élément scellé 10 et de la structure 14 ne peut être directement reportée sur la droite D (R0, a) illustrant cette loi L de variation entre la raideur dynamique et l'effort statique, car cette dernière provient d'essais effectués uniquement sur des éléments scellés seuls et les essais dynamiques sont effectués in situ, sur l'élément scellé 10 et la structure 14. Il convient donc de corriger cette raideur dynamique R pour pouvoir déterminer l'effort de traction auquel est soumis l'élément scellé 10 seul pour maintenir la structure 14 contre le support 22.

[0042]   A ce stade, deux modes opératoires peuvent être effectués selon la nature des paramètres connus.

[0043]   En effet, le paramètre traduisant l'influence de la structure 14 sur la raideur dynamique R de l'élément scellé 10 peut être déterminé de deux manières : à partir d'essais effectués sur la structure 14 seule ou bien à partir d'essais effectués sur l'élément scellé et la structure 14. La détermination de l'effort de traction auquel est soumis l'élément scellé 10 différera alors selon le paramètre déterminé, comme on le verra de manière plus détaillée ci-après.

[0044]   Selon un premier mode opératoire, on détermine préalablement une raideur dynamique intrinsèque Rs de la structure 14 seule.

[0045]   Cette raideur dynamique intrinsèque Rs de la structure 14 peut être déterminée à partir de banques de données connues pour des éléments du type de l'élément de structure 14 et de la connaissance de l'épaisseur de l'élément de structure 14 utilisé in situ.

[0046]   Lorsque la raideur dynamique intrinsèque Rs de la structure 14 n'est pas connue, mais que les éléments scellés ne sont pas encore mis en place sur site, on peut effectuer des essais dynamiques sur des structures 14 (sans élément scellé) en effectuant les étapes b) à d) précitées sur ces éléments de structure 14 seuls et ainsi déterminer la raideur dynamique Rs en relevant l'inverse de la pente à l'origine (à 2π près), comme exposé dans l'étape d), cette raideur Rs étant ensuite gardée en mémoire pour, le moment venu, être utilisée pour corriger des mesures selon l'invention.

[0047]   Un essai dynamique est par ailleurs réalisé sur un élément scellé 10 mis en place contre une structure 14 et contraint en traction, pour déterminer la raideur dynamique R de l'élément scellé 10 et de la structure 14, en appliquant les étapes b) à d).

[0048]   Dans ce mode opératoire, la raideur dynamique Rs de la structure 14 étant connue, elle représente directement le paramètre de correction Ps. Ainsi, il suffit de corriger, à l'aide de moyens 38 pour déterminer la raideur dynamique corrigée Rcs, la racine carrée de la raideur dynamique R de l'élément scellé 10 et de la structure 14 en faisant la différence avec la racine carrée de cette première raideur dynamique Rs et de reporter la raideur dynamique corrigée Rcs correspondante sur un graphe G2 représenté sur la figure 6. On reporte aussi sur ce graphe G2, la loi de variation L correspondant à la droite D (R0, a).

[0049]   En fait,

$$\begin{cases} Rcs^{\frac{1}{2}} = R^{\frac{1}{2}} - Ps^{\frac{1}{2}} \\ Ps = Rs \end{cases}.$$

[0050]   L'effort dans l'élément scellé 10 est obtenu en appliquant la loi de variation à cette raideur dynamique corrigée Rcs. En l'espèce, il convient de lire sur la droite D (R0, a) du graphe G2 de la figure 6, l'effort F1 correspondant à la valeur de la raideur dynamique corrigée Rcs qui est reportée en ordonnée. Le dispositif comporte à cet effet des moyens

EP 1 761 750 B1

40 pour appliquer la loi de variation L et déterminer l'effort F1.

[0051]   Selon un autre mode opératoire, lorsqu'il n'est pas possible de connaître la raideur dynamique intrinsèque Rs de la structure 14 seule, en particulier lorsque la détermination de l'effort de traction se fait sur des sites sur lesquels les éléments scellés 10 sont déjà mis en place, on détermine préalablement une raideur dynamique R1 de l'élément scellé 10 mis en place contre la structure 14 et non contraint, pour déduire ensuite l'influence de la structure.

[0052]   Pour ce faire on effectue les étapes b) à d) précitées, en réalisant des essais dynamiques sur un élément scellé 10 mis en place contre une structure 14 et non encore contraint en traction.

[0053]   On reporte cette valeur de raideur dynamique R1 de l'élément scellé 10 avec la structure 14 sur un graphe G3 représenté sur la figure 7 et on construit la droite D (R1, a) en appliquant la loi L ayant pour origine cette valeur de la raideur dynamique R1 pour l'élément scellé 10 et la structure 14. Cette valeur de raideur dynamique R1 pour l'élément scellé 10 et la structure 14 représente la raideur dynamique de l'élément scellé et l'influence qu'a la structure sur ce dernier.

[0054]   De manière analogue à celle décrite pour le premier mode opératoire, un essai dynamique est par ailleurs réalisé sur un élément scellé 10 mis en place contre une structure 14 et contraint en traction, pour déterminer la raideur dynamique R de l'élément scellé 10 et de la structure 14, en appliquant les étapes b) à d).

[0055]   Cependant dans ce mode opératoire, la raideur dynamique Rs de la structure seule 14 n'étant pas connue, il faut déterminer un autre paramètre de correction P traduisant l'influence de la structure sur la raideur dynamique R de l'élément scellé 10 et de la structure 14.

[0056]   Ainsi, à l'aide des moyens 38 pour déterminer la raideur dynamique corrigée Rc, en faisant la différence entre les racines carrées de la raideur dynamique R1 de l'élément scellé 10 avec la structure 14 et de la raideur dynamique intrinsèque R0 de l'élément scellé sans la structure, on obtient un paramètre de correction P.

[0057]   En fait,

$$\begin{cases} Rc^{\frac{1}{2}} = R^{\frac{1}{2}} - P \\ P = R1^{\frac{1}{2}} - R0^{\frac{1}{2}} \end{cases}.$$

[0058]   Cependant, ce paramètre de correction P ne représente pas directement la raideur dynamique de la structure seule 14, puisqu'il découle de mesures faites sur l'élément scellé 10 et la structure 14, l'un interagissant avec l'autre.

[0059]   En conséquence, il convient de reporter la raideur dynamique corrigée Rc non pas sur la droite D (R0, a) correspondant à l'application de la loi L pour l'élément scellé seul (non encore contraint), mais sur la droite D (R1, a) correspondant à l'application de la loi L pour l'élément scellé 10 (non encore contraint) maintenant la structure 14.

[0060]   L'effort dans l'élément scellé 10 est obtenu en appliquant la loi de variation à cette raideur dynamique corrigée Rc, à l'aide des moyens 40 pour appliquer la loi de variation L et déterminer l'effort F2. En l'espèce, il convient de lire sur la droite D (R1, a) du graphe G3 de la figure 7, l'effort F2 correspondant à la valeur de la raideur dynamique corrigée Rc qui est reportée en ordonnée.

[0061]   Il est à noter que quelque soit le mode opératoire retenu, les valeurs des efforts de traction F1 et F2 obtenus restent proches l'un de l'autre. En conséquence, F1 ou F2 représente bien l'effort de traction auquel est soumis l'élément scellé 10.

Revendications

1. Procédé de détermination de l'effort de traction (F1, F2) auquel est soumis un élément scellé (10) maintenant une structure (14) contre un support (22), l'élément scellé (10) étant contraint en traction, comprenant les étapes suivantes:

      a) on fournit une loi de variation (L) entre la raideur dynamique de l'élément scellé (10) et un effort statique de traction (F) auquel ce dernier (10) est soumis, F)
      b) on soumet l'élément scellé (10) à un impact de force déterminée (Fd), pour générer une vibration de l'élément scellé (10) et de la structure (14),
      c) on relève la réponse vibratoire (32, 32A, 32B, 34) de l'élément scellé (10) et de la structure (14) mis en vibration,
      d) on détermine la raideur dynamique (R) de l'élément scellé (10) et de la structure (14) à partir de la réponse vibratoire (32, 32A, 32B, 34),
      e) on corrige la raideur dynamique (R) de l'élément scellé (10) et de la structure (14) d'un paramètre (P, Ps) traduisant l'influence de la structure (14) sur cette dernière raideur dynamique (R), et

EP 1 761 750 B1

f) on applique la loi de variation (L) à cette raideur dynamique corrigée (Rc, Rcs) pour obtenir l'effort (F1, F2) dans l'élément scellé (10).

2. Procédé selon la revendication 1, **caractérisé en ce que** l'impact (Fd) est réalisé sur la tête (12A) de l'élément scellé (10).

3. Procédé selon la revendication 1, **caractérisé en ce que** l'impact (Fd) est réalisé sur une plaque d'appui (16) de l'élément scellé (10).

4. Procédé selon l'une quelconque des revendications précédentes, **caractérisé en ce que** la loi (L) définit une droite d'équation :

$$(Rd)^{\frac{1}{2}} = \underline{a} \times F + (Ri)^{\frac{1}{2}}$$

où Rd représente la raideur dynamique de l'élément testé,
Ri représente la raideur de l'élément testé,
F représente l'effort dans l'élément testé, et
$\underline{a}$ dépend de l'élément testé.

5. Procédé selon l'une quelconque des revendications 1 à 4, **caractérisé en ce que** :

- on fournit une raideur dynamique intrinsèque de la structure (Rs) qui correspond au paramètre (Ps) traduisant l'influence de la structure (14) sur la raideur dynamique (R) de l'élément scellé (10) et de la structure (14),
- on fournit une raideur dynamique intrinsèque (R0) de l'élément scellé (10), et
- on applique sur une courbe (D (R0, $\underline{a}$)) représentant la raideur dynamique de l'élément scellé (10), la raideur dynamique corrigée (Rcs) par ledit paramètre (Ps), pour obtenir l'effort dans l'élément scellé (10).

6. Procédé selon l'une quelconque des revendications 1 à 4, **caractérisé en ce que** :

- on fournit une raideur dynamique (R1) de l'élément scellé (10) avec la structure (14),
- on fournit une raideur dynamique intrinsèque (R0) de l'élément scellé (10),
- on construit une courbe (D (R1, $\underline{a}$)) représentant la raideur dynamique de l'élément scellé (10) et de la structure (14) à partir de l'application de la loi de variation (L) à la raideur dynamique (R1) de l'élément scellé (10) avec la structure (14),
- on détermine le paramètre (P) traduisant l'influence de la structure (14) sur la raideur dynamique (R) de l'élément scellé (10) et de la structure (14), en faisant la différence entre la raideur dynamique (R1) de l'élément scellé (10) avec la structure (14) et une raideur dynamique intrinsèque (R0) de l'élément scellé (10) sans la structure (14), et
- on applique sur la courbe (D (R1, $\underline{a}$)) représentant la raideur dynamique (R1) de l'élément scellé (10) et de la structure (14), la raideur dynamique corrigée (Rc) par ledit paramètre (P), pour obtenir l'effort dans l'élément scellé (10).

7. Procédé selon la revendication 5 ou 6, **caractérisé en ce qu'**on évalue la raideur dynamique intrinsèque (R0) de l'élément scellé (10) seul en effectuant les étapes b) à d) sur un élément scellé (10) seul et non encore contraint en traction.

8. Procédé selon la revendication 5, **caractérisé en ce qu'**on évalue la raideur dynamique intrinsèque (Rs) de la structure (14) en effectuant les étapes b) à d) sur une structure sans élément scellé (10).

9. Procédé selon la revendication 6, **caractérisé en ce qu'**on évalue la raideur dynamique (R1) de l'élément scellé (10) avec la structure (14) en effectuant les étapes b) à d) sur un élément scellé (10) non contraint.

10. Dispositif de détermination de l'effort de traction (F1, F2) auquel est soumis un élément scellé (10) d'une structure (14) contre un support (22), l'élément scellé (10) étant scellé et contraint en traction par rapport au support (22), comportant:

7

EP 1 761 750 B1

- une mémoire (M) contenant une loi de variation (L) entre la raideur dynamique de l'élément scellé (10) et un effort statique (F) de traction auquel ce dernier (10) est soumis,
- des moyens (26) pour soumettre l'élément scellé (10) à un impact de force déterminée (Fd) pour générer une vibration de l'élément scellé (10) et de la structure.
5
- des moyens (28) pour relever la réponse vibratoire (32, 32A, 32B, 34) de l'élément scellé (10) et de la structure (14) mis en vibration,
- des moyens (36) pour déterminer la raideur dynamique (R) de l'élément scellé (10) et de la structure (14) à partir de la réponse vibratoire (32, 32A, 32B, 34),
- des moyens (36) pour déterminer la raideur dynamique (Rc, Rcs) de l'élément scellé (10) permettant de
10
corriger la raideur dynamique (R) de l'élément scellé (10) et de la structure (14) d'un paramètre (P, Ps) traduisant l'influence de la structure (14) sur cette dernière raideur dynamique (R), et
- des moyens (40) pour appliquer la loi de variation (L) à cette raideur dynamique corrigée (Rc, Rcs) pour obtenir l'effort (F₁, F₂) dans l'élément scellé (10).

15   **11.** Dispositif selon la revendication précédente, **caractérisé en ce qu'**il comporte des moyens de traitement mathématiques (T) destinés à traiter la réponse vibratoire de l'élément scellé (10) et de la structure (14).

Claims
20
1. Method for determining the tensile stress (F₁, F₂) exerted on a sealed element (10) holding a structure (14) against a support (22), the sealed element (10) being subject to tensile stress, comprising the following steps:

a) providing a variation law (L) between the dynamic stiffness of the sealed element (10) and a static tensile
25   stress (F) exerted on the latter (10).
b) subjecting the sealed element (10) to an impact of a given force (Fd), in order to generate vibration in the sealed element (10) and the structure (14),
c) recording the vibratory response (32, 32A, 32B, 34) of the sealed element (10) and the structure (14) set in vibration,
30   d) determining the dynamic stiffness (R) of the sealed element (10) and the structure (14) from the vibratory response (32, 32A, 32B, 34),
e) correcting the dynamic stiffness (R) of the sealed element (10) and the structure (14) of a parameter (P, Ps) expressing the influence of the structure (14) on the latter dynamic stiffness (R), and
f) applying the variation law (L) to this corrected dynamic stiffness (Rc, Rcs) in order to achieve stress (F₁, F₂)
35   in the sealed element (10).

2. Method according to Claim 1, **characterised in that** the impact (Fd) is produced on the head (12A) of the sealed element (10).

40 **3.** Method according to Claim 1, **characterised in that** the impact (Fd) is produced on a base plate (16) of the sealed element (10).

4. Method according to any one of the preceding claims, **characterised in that** the law (L) defines a straight line with the equation:
45

$$(Rd)^{\frac{1}{2}} = \underline{a} \times F + (Ri)^{\frac{1}{2}}$$

50
where Rd represents the dynamic stiffness of the element tested,
Ri represents the stiffness of the element tested,
F represents the stress in the element tested, and
$\underline{a}$ depends on the element being tested.
55
5. Method according to any one of Claims 1 to 4, **characterised in that:**

- an intrinsic dynamic stiffness of the structure (Rs) is provided, which corresponds to the parameter (Ps)

8

EP 1 761 750 B1

expressing the influence of the structure (14) on the dynamic stiffness (R) of the sealed element (10) and the structure (14).
- an intrinsic dynamic stiffness (R0) of the sealed element (10) is provided, and
- the dynamic stiffness (Rcs) corrected by said parameter (Ps) is applied to a curve (D (R0, a)) representing the dynamic stiffness of the sealed element (10), in order to achieve stress in the sealed element (10).

6. Method according to any one of Claims 1 to 4, **characterised in that:**

   - dynamic stiffness (R1) of the sealed element (10) together with the structure (14) is provided,
   - intrinsic dynamic stiffness (R0) of the sealed element (10) is provided,
   - a curve (D (R1, a)) is constructed, representing the dynamic stiffness of the sealed element (10) and the structure (14) by applying the variation law (L) to the dynamic stiffness (R1) of the sealed element (10) together with the structure (14),
   - the parameter (P) expressing the influence of the structure (14) on the dynamic stiffness (R) of the sealed element (10) and the structure (14) is determined, by making the difference between the dynamic stiffness (R1) of the sealed element (10) together with the structure (14) and the intrinsic dynamic stiffness (R0) of the sealed element (10) without the structure (14), and
   - the dynamic stiffness (Rc) corrected by said parameter (P) is applied to the curve (D (R1, a)) representing the dynamic stiffness (R1) of the sealed element (10) and the structure (14), in order to achieve stress in the sealed element (10).

7. Method according to Claim 5 or 6, **characterised in that** the intrinsic dynamic stiffness (R0) of the sealed element (10) is evaluated by performing the steps b) to d) on a sealed element (10) on its own and not yet subject to tensile stress.

8. Method according to Claim 5, **characterised in that** the intrinsic dynamic stiffness (Rs) of the structure (14) is evaluated by performing the steps b) to d) on a structure without a sealed element (10).

9. Method according to Claim 6, **characterised in that** the dynamic stiffness (R1) of the sealed element (10) together with the structure (14) is evaluated by performing the steps b) to d) on a non-stressed sealed element (10).

10. Device for determining the tensile stress (F_1, F_2) exerted on a sealed element (10) of a structure (14) against a support (22), the sealed element (10) being sealed and subject to tensile stress with respect to the support (22), including:

   - a memory (M) containing a variation law (L) between the dynamic stiffness of the sealed element (10) and a static tensile stress (F) exerted on the latter (10),
   - means (26) for subjecting the sealed element (10) to an impact of a given force (Fd) in order to generate vibration in the sealed element (10) and the structure,
   - means (28) for recording the vibratory response (32, 32A, 32B, 34) of the sealed element (10) and the structure (14) set in vibration,
   - means (36) for determining the dynamic stiffness (R) of the sealed element (10) and the structure (14) from the vibratory response (32, 32A, 32B, 34),
   - means (38) for determining the dynamic stiffness (Rc, Rcs) of the sealed element (10) making it possible to correct the dynamic stiffness (R) of the sealed element (10) and the structure (14) of a parameter (P, Ps) expressing the influence of the structure (14) on the latter dynamic stiffness (R), and
   - means (40) for applying the variation law (L) to this corrected dynamic stiffness (Rc, Rcs) in order to achieve stress (F_1, F_2) in the sealed element (10).

11. Device according to the preceding claim, **characterised in that** it comprises mathematical processing means (T) intended to process the vibratory response of the sealed element (10) and the structure (14).

Patentansprüche

1. Verfahren zum Bestimmen der Zugkraft (F1, F2), der ein vergossenes Element (10), das einen Aufbau (14) gegen einen Träger (22) hält, ausgesetzt ist,
wobei das vergossene Element (10) einer Zugbeanspruchung ausgesetzt ist, umfassend die folgenden Schritte:

a) man stellt ein Variationsgesetz (L) zwischen der dynamischen Steifigkeit des vergossenen Elements (10) und einer statischen Zugkraft (F) bereit, welcher das Elements (10) ausgesetzt ist,

b) man unterwirft das vergossene Element (10) einem Stoß mit einer bestimmten Kraft (Fd), um eine Schwingung des vergossenen Elements (10) und des Aufbaus (14) zu erzeugen,

c) man mißt die Schwingungsreaktion (32, 32A, 32B, 34) des vergossenen Elements (10) und des Aufbaus (14), die in Schwingung versetzt wurden,

d) man bestimmt die dynamische Steifigkeit (R) des vergossenen Elements (10) des Aufbaus (14) ausgehend von der Schwingungsreaktion (32, 32A, 32B, 34),

e) man korrigiert die dynamische Steifigkeit (R) des vergossenen Elements (10) und des Aufbaus (14) um einen Parameter (P, Ps), der den Einfluß des Aufbaus (14) auf diese letztere dynamische Steifigkeit (R) wiedergibt, und

f) man wendet das Variationsgesetz (L) auf diese korrigierte dynamische Steifigkeit (Rc, Rcs) an, um die Kraft (F1, F2) in dem vergossenen Element (10) zu erzielen.

2. Verfahren nach Anspruch 1, **dadurch gekennzeichnet, daß** der Stoß (Fd) auf dem Kopf (12A) des vergossenen Elements (10) ausgeführt wird.

3. Verfahren nach Anspruch 1, **dadurch gekennzeichnet, daß** der Stoß (Fd) auf einer Auflageplatte (16) des vergossenen Elements (10) ausgeführt wird.

4. Verfahren nach einem der vorhergehenden Ansprüche, **dadurch gekennzeichnet, daß** das Gesetz (L) eine Gerade mit der folgenden Gleichung definiert:

$$Rd^{1/2} = \underline{a} \times F + (Ri)^{1/2}$$

wobei Rd die dynamische Steifigkeit des getesteten Elements darstellt,
Ri die Steifigkeit des getesteten Elements darstellt,
F die Kraft in dem getesteten Element darstellt und
$\underline{a}$ von dem getesteten Element abhängt.

5. Verfahren nach einem der Ansprüche 1 bis 4, **dadurch gekennzeichnet, daß**:

- man eine intrinsische dynamische Steifigkeit des Aufbaus (Rs), die dem Parameter (Ps) entspricht, der den Einfluß des Aufbaus (14) auf die dynamische Steifigkeit (R) des vergossenen Elements (10) und des Aufbaus (14) wiedergibt, bereitstellt,
- man eine intrinsische dynamische Steifigkeit (R0) des vergossenen Elements (10) bereitstellt, und
- man auf eine Kurve (D (R0, $\underline{a}$)), die die dynamische Steifigkeit des vergossenen Elements (10) darstellt, die mit dem Parameter (Ps) korrigierte dynamische Steifigkeit (Rcs) anwendet, um die Kraft in dem vergossenen Element (10) zu erzielen.

6. Verfahren nach einem der Ansprüche 1 bis 4, **dadurch gekennzeichnet, daß**

- man eine dynamische Steifigkeit (R1) des vergossenen Elements (10) mit dem Aufbau (14) bereitstellt,
- man eine intrinsische dynamische Steifigkeit (R0) des vergossenen Elements (10) bereitstellt,
- man eine Kurve (D (R1, $\underline{a}$)) erstellt, die die dynamische Steifigkeit des vergossenen Elements (10) und des Aufbaus (14) ausgehend von dem Anwenden des Variationsgesetzes (L) auf die dynamische Steifigkeit (R1) des vergossenen Elements (10) mit dem Aufbau (14) darstellt,
- man den Parameter (P), der den Einfluß des Aufbaus (14) auf die dynamische Steifigkeit (R) des vergossenen Elements (10) und des Aufbaus (14) wiedergibt, bestimmt, indem man den Unterschied zwischen der dynamischen Steifigkeit des vergossenen Elements (10) mit dem Aufbau (14) und einer intrinsischen dynamischen Steifigkeit (R0) des vergossenen Elements (10) ohne den Aufbau (14) bestimmt, und
- man auf die Kurve (D (R1, $\underline{a}$)), die die dynamische Steifigkeit (R1) des vergossenen Elements (10) und des Aufbaus (14) darstellt, die mit dem Parameter (P) korrigierte dynamische Steifigkeit (Rc) anwendet, um die Kraft in dem vergossenen Element (10) zu erzielen.

7. Verfahren nach Anspruch 5 oder 6, **dadurch gekennzeichnet, daß** man die intrinsische dynamische Steifigkeit

EP 1 761 750 B1

(R0) des vergossenen Elements (10) allein **dadurch** bewertet, indem man die Schritte b) bis d) an einem vergossenen Element (10) allein, das noch keiner Zugbelastung ausgesetzt ist, durchführt.

8.   Verfahren nach Anspruch 5, **dadurch gekennzeichnet, daß** man die intrinsische dynamische Steifigkeit (Rs) des Aufbaus (14) bewertet, indem man die Schritte b) bis d) auf einem Aufbau ohne vergossenes Element (10) durchführt.

9.   Verfahren nach Anspruch 6, **dadurch gekennzeichnet, daß** man die dynamische Steifigkeit (R1) des vergossenen Elements (10) mit dem Aufbau (14) bewertet, indem man die Schritte b) bis d) an einem vergossenen Element (10) ohne Belastung ausführt.

10. Vorrichtung zum Bestimmen der Zugkraft (F1, F2), der ein vergossenes Element (10) eines Aufbaus (14) gegen einen Träger (22) ausgesetzt ist, wobei das vergossene Element (10) vergossen und einer Zugbeanspruchung relativ zu dem Träger (22) ausgesetzt ist, umfassend:

- einen Speicher (M), der das Variationsgesetz (L) zwischen der dynamischen Steifigkeit des vergossenen Elements (10) und einer statischen Zugkraft (F), der dieses Letztere (10) unterworfen wird, enthält,
- Mittel (26), um das vergossene Element (10) einem Stoß mit bestimmter Kraft (Fd) zu unterwerfen, um eine Schwingung des vergossenen Elements (10) und des Aufbaus zu erzeugen,
- Mittel (28) zum Messen der Schwingungsreaktion (32, 32A, 32B, 34) des vergossenen Elements (10) und des Aufbaus (14), die in Schwingung versetzt wurden,
- Mittel (36) zum Bestimmen der dynamischen Steifigkeit (R) des vergossenen Elements (10) und des Aufbaus (14) ausgehend von der Schwingungsreaktion (32, 32A, 32B, 34),
- Mittel (38), um die dynamische Steifigkeit (Rc, Rcs) des vergossenen Elements (10) zu bestimmen, die es erlauben, die dynamische Steifigkeit (R) des vergossenen Elements (10) und des Aufbaus (14) mit einem Parameter (P, Ps) zu korrigieren, der den Einfluß der Struktur (14) auf diese letztere dynamische Steifigkeit (R) wiedergibt, und
- Mittel (40) zum Anwenden des Variationsgesetzes (L) auf diese korrigierte dynamische Steifigkeit (Rc, Rcs), um die Kraft (F1, F2) in dem vergossenen Element (10) zu erzielen.

11. Vorrichtung nach dem vorhergehenden Anspruch, **dadurch gekennzeichnet, daß** sie mathematische Verarbeitungsmittel (T) aufweist, die dazu bestimmt sind, die Schwingungsreaktion des vergossenen Elements (10) und des Aufbaus (14) zu verarbeiten.

EP 1 761 750 B1

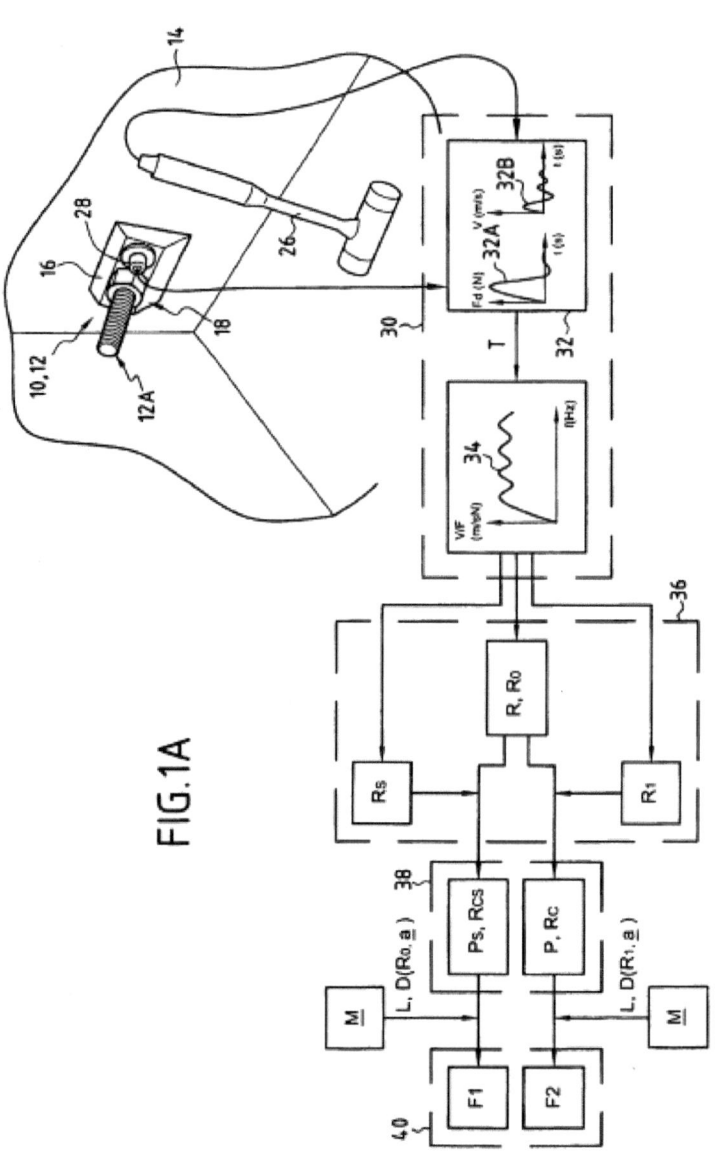

FIG.1A

EP 1 761 750 B1

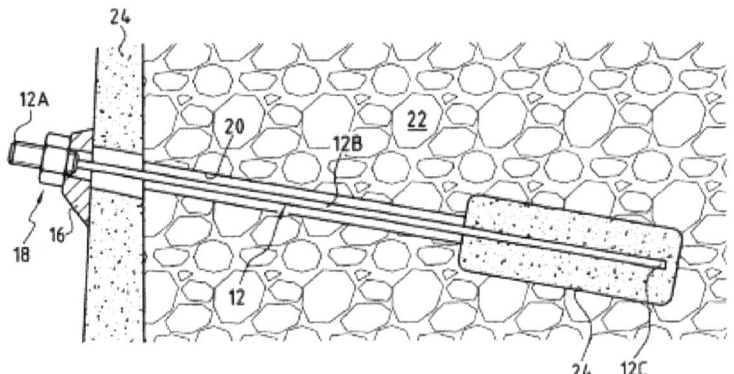

FIG.1B

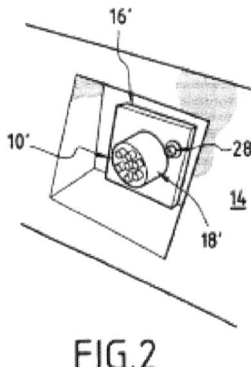

FIG.2

13

EP 1 761 750 B1

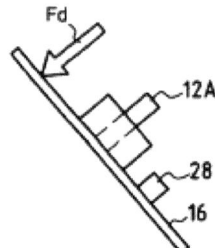

FIG.3

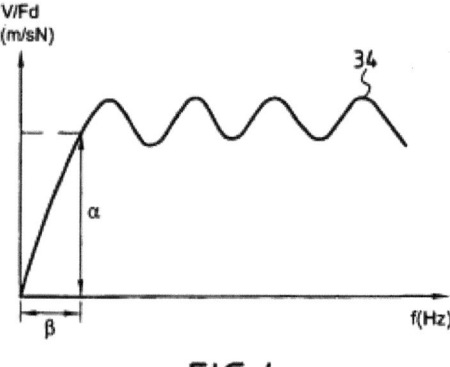

FIG.4

14

EP 1 761 750 B1

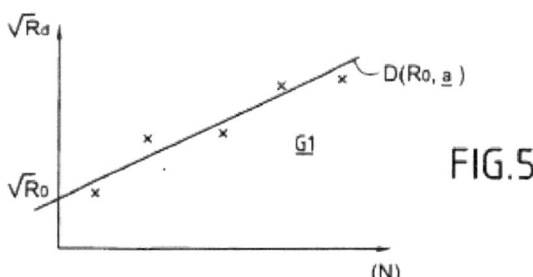

FIG.5

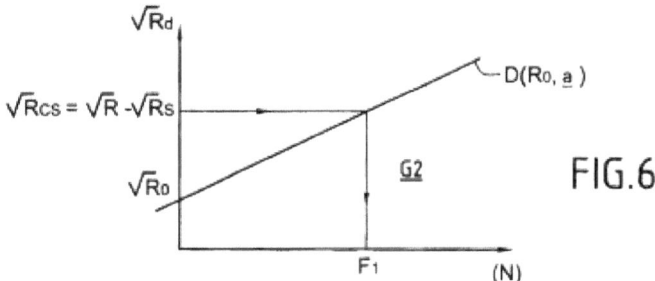

FIG.6

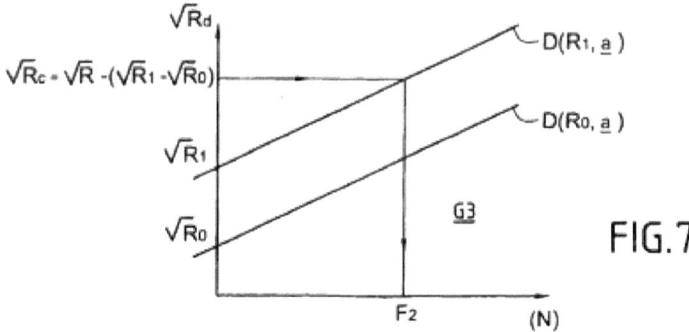

FIG.7

EP 1 761 750 B1

RÉFÉRENCES CITÉES DANS LA DESCRIPTION

Documents brevets cités dans la description

- US 4979125 A **[0005]**
- EP 0438347 A **[0005]**
- WO 9953282 A **[0005]**
- . GB 1446661 A **[0005]**

Appendix B

JJ. RINCENT—Method and device for determining the tensile stress exerted on a sealed element. **Patent OMPI n° WO 2006/010830 A1, 2 February 2006.**

© The Editor(s) (if applicable) and The Author(s) 2024

J.-J. H. Rincent, *Ground Anchors*, https://doi.org/10.1007/978-981-97-4414-5

(12) DEMANDE INTERNATIONALE PUBLIÉE EN VERTU DU TRAITÉ DE COOPÉRATION
EN MATIÈRE DE BREVETS (PCT)

(19) Organisation Mondiale de la Propriété
Intellectuelle
Bureau international

(43) Date de la publication internationale
2 février 2006 (02.02.2006)   PCT

(10) Numéro de publication internationale
WO 2006/010830 A1

(51) Classification internationale des brevets⁷ : G01L 1/10, G01N 3/34

(21) Numéro de la demande internationale : PCT/FR2005/001597

(22) Date de dépôt international : 24 juin 2005 (24.06.2005)

(25) Langue de dépôt : français

(26) Langue de publication : français

(30) Données relatives à la priorité :
0406969   25 juin 2004 (25.06.2004)   FR

(71) Déposant (pour tous les États désignés sauf US) : RIN-CENT BTP SERVICES [FR/FR]; 39 Rue Michel Ange, Parc Elysée, F-91026 EVRY Cedex (FR).

(72) Inventeur; et

(75) Inventeur/Déposant (pour US seulement) : RINCENT, Jean-Jacques [FR/FR]; c/o RINCENT BTP Services, 39

Rue Michel Ange - Parc Elysée, F-91026 EVRY Cedex (FR).

(74) Mandataires : INTES, Didier etc.; Cabinet Beau de Loménie, 158 Rue de l'Université, F-75340 PARIS (FR).

(81) États désignés (sauf indication contraire, pour tout titre de protection nationale disponible) : AE, AG, AL, AM, AT, AU, AZ, BA, BB, BG, BR, BW, BY, BZ, CA, CH, CN, CO, CR, CU, CZ, DE, DK, DM, DZ, EC, EE, EG, ES, FI, GB, GD, GE, GH, GM, HR, HU, ID, IL, IN, IS, JP, KE, KG, KM, KP, KR, KZ, LC, LK, LR, LS, LT, LU, LV, MA, MD, MG, MK, MN, MW, MX, MZ, NA, NG, NI, NO, NZ, OM, PG, PH, PL, PT, RO, RU, SC, SD, SE, SG, SK, SL, SM, SY, TJ, TM, TN, TR, TT, TZ, UA, UG, US, UZ, VC, VN, YU, ZA, ZM, ZW.

(84) États désignés (sauf indication contraire, pour tout titre de protection régionale disponible) : ARIPO (BW, GH,

[Suite sur la page suivante]

(54) Title: METHOD AND DEVICE FOR DETERMINING THE TENSILE STRESS EXERTED ON A SEALED ELEMENT

(54) Titre : PROCEDE ET DISPOSITIF DE DETERMINATION DE L'EFFORT DE TRACTION AUQUEL EST SOUMIS UN ELEMENT SCELLE

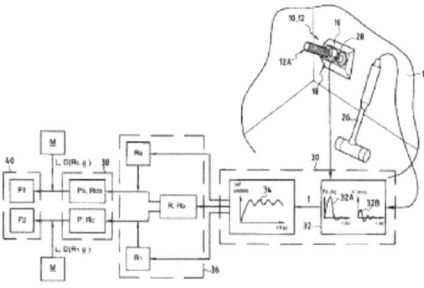

(57) Abstract: The invention concerns a method and device for determining the tensile stress (F1, F2) exerted on a sealed element (10) maintaining a structure (14), the sealed element (10) being subjected to tensile stress. The invention is characterized in that it consists in: a) providing a variation law (L) between the dynamic stiffness of the sealed element (10) and a static stress exerted on the latter (10); b) subjecting the sealed element (10) to an impact of determined force, to generate a vibration; c) recording the vibratory response (32, 32A, 32B, 34) of the sealed element (10) and of the vibrated structure (14); d) determining the dynamic stiffness (R) of the sealed element (10) and the structure (14); e) correcting the dynamic stiffness (R) with a parameter (P, Ps) expressing the influence of the structure (14); and f) applying the variation law (L) to said corrected stiffness (R, Rcs) to obtain the stress (F1, F2) in the sealed element (10).

(57) Abrégé : L'invention concerne un procédé et un dispositif de détermination de l'effort de traction (F1, F2) auquel est soumis un élément scellé (10) maintenant une structure (14), l'élément scellé (10) étant contraint en traction. Selon l'invention : a) on fournit une loi de variation (L)

[Suite sur la page suivante]

WO 2006/010830 A1

GM, KE, LS, MW, MZ, NA, SD, SL, SZ, TZ, UG, ZM, ZW), eurasien (AM, AZ, BY, KG, KZ, MD, RU, TJ, TM), européen (AT, BE, BG, CH, CY, CZ, DE, DK, EE, ES, FI, FR, GB, GR, HU, IE, IS, IT, LT, LU, MC, NL, PL, PT, RO, SE, SI, SK, TR), OAPI (BF, BJ, CF, CG, CI, CM, GA, GN, GQ, GW, ML, MR, NE, SN, TD, TG).

Publiée :
— avec rapport de recherche internationale

En ce qui concerne les codes à deux lettres et autres abréviations, se référer aux "Notes explicatives relatives aux codes et abréviations" figurant au début de chaque numéro ordinaire de la Gazette du PCT.

Appendix C

ISSN 0335-3931

norme française

NF P 94-160-4
Mars 1994

Indice de classement : P 94-160-4

Sols : reconnaissance et essais

Auscultation d'un élément de fondation

Partie 4 : Méthode par impédance

E : Soil : investigation and testing — Auscultation of buried work —
Part 4 : Impedance test
D : Bodenerkündung und Prüfungen — Prüfung eines eingegrabenen
Bauwerkes — Teil 4 : Impedanzverfahren

Norme française homologuée par décision du Directeur Général de l'AFNOR
le 5 février 1994 pour prendre effet le 5 mars 1994.

correspondance   A la date de publication du présent document, il n'existe pas de travaux internationaux ou européens sur ce sujet.

analyse   Le présent document traite de l'essai d'auscultation d'un pieu en béton armé
ou non par la méthode par impédance. Il définit les termes employés et les
paramètres mesurés, spécifie les caractéristiques de l'appareillage, fixe le
mode opératoire de l'essai et précise les résultats à présenter.

descripteurs   **Thésaurus International Technique** : sol, fondation, pieu de fondation, pieu en
béton, examen, propagation des ondes, essai, impédance.

modifications

corrections

édité et diffusée par l'association française de normalisation (afnor), tour europe cedex 7 92049 paris la défense — tél. : (1) 42 91 55 55

AFNOR 1994   © AFNOR 1994   1ᵉʳ tirage 94-03

Sols : reconnaissance et essais

BNSR - **SRE**

Membres de la commission de normalisation

Président : M PAREZ

Secrétariat : M BIGOT — Laboratoire Régional de l'Est Parisien

M	AMAR	Laboratoire Central des Ponts et Chaussées
M	BARNOUD	Union Syndicale Géotechnique
M	BLONDEAU	Comité Professionnel de la Prévention et du Contrôle Technique
M	CASSAN	FONDASOL
M	CHAILLOT	SNCF — Direction de l'Equipement
MME	DAURELLE	AFNOR
M	DEBATTISTA	EDF — TEGG — DGC
M	DORÉ	MECASOL
M	GONIN	SIMECSOL
M	LEGENDRE	Sondage, Forage et Fondations Spéciales — Syndicat National des Entrepreneurs
M	PAREZ	SOL — ESSAIS
M	PHILIPPONNAT	SOPENA
M	RINCENT	Centre Expérimental de Recherches et d'Etudes du Bâtiment et des Travaux Publics

Ont participé en tant qu'experts :

M	GENDRE	Centre Expérimental de Recherches et d'Etudes du Bâtiment et des Travaux Publics
M	LELIEVRE	Laboratoire Régional de Rouen — CETE Normandie Centre
M	PINCENT	SIMECSOL

Appendix D

ISSN 0335-3931

NF P 94-153
Décembre 1993

norme française

Indice de classement : P 94-153

Sols : reconnaissance et essais

Essai statique de tirant d'ancrage

E : Soils investigation and testing — Tension rod test
D : Bodenerkundung und Prüfungen — Statische Prüfung des Zugankers

Norme française homologuée par décision du Directeur Général de l'AFNOR le 20 novembre 1993 pour prendre effet le 20 décembre 1993.
Remplace la norme de même indice, de novembre 1991.

correspondance A la date de publication du présent document, il n'existe pas de travaux internationaux ou européens en cours sur le même sujet.

analyse Le présent document fixe la terminologie, l'appareillage, le mode opératoire et la méthode de calcul des différents paramètres déduits des essais de tirant d'ancrage.

descripteurs **Thésaurus International Technique** : sol, fondation, essai statique, essai d'arrachement, essai de conformité, contrôle, mode opératoire, appareillage, calcul.

modifications Par rapport à la précédente édition, changement des classes d'essai, de la durée des paliers de chargement pour le tirant ER_2, des critères de rupture et de charge de fluage, du calcul des pentes des droites de stabilisation, de la loi de chargement lors d'un essai de contrôle.

corrections

éditée et diffusée par l'association française de normalisation (afnor), tour europe cedex 7 92049 paris la défense — tél. : (1) 42 91 55 55

AFNOR 1993   © AFNOR 1993   1er tirage 93-12

Sols, reconnaissance et essais   BNSR SRE

Membres de la commission de normalisation

Président : M PAREZ

Secrétariat : M BIGOT — LABORATOIRE REGIONAL DE L'EST PARISIEN

M	AMAR	LABORATOIRE CENTRAL DES PONTS ET CHAUSSEES
M	BARNOUD	UNION SYNDICALE GEOTECHNIQUE
M	BLONDEAU	COMITE PROFESSIONNEL DE LA PREVENTION ET DU CONTROLE TECHNIQUE
M	CASSAN	FONDASOL
M	CHAILLOT	SNCF — DIRECTION DE L'EQUIPEMENT
M	DEBATTISTA	EDF — TEGG — DGC
M	DORE	MECASOL
MME	FERNANDEZ	AFNOR
M	GONIN	SIMECSOL
M	LEGENDRE	SONDAGE, FORAGE ET FONDATIONS SPECIALES — SYNDICAT NATIONAL DES ENTREPRENEURS
M	PAREZ	SOLS — ESSAIS
M	PHILIPPONNAT	SOPENA
M	RINCENT	CENTRE EXPERIMENTAL DE RECHERCHES ET D'ETUDES DU BATIMENT ET DES TRAVAUX PUBLICS (CEBTP)

Ont participé en tant qu'experts :
Les membres du groupe Recommandations TA 93 :

Président : M HABIB

Secrétariat : M LOGEAIS — INGENIEUR CONSEIL

M	BERTHELOT	VERITAS
M	BIGOT	LREP
M	BOUCHERIE	SOCOTEC
M	BRIN	CEP
M	BUSTAMANTE	LCPC
M	CLEMENT	SIF BACHY
M	DUPEUPLE	SIF BACHY
M	HABIB	GROUPEMENT POUR L'ETUDE DES STRUCTURES SOUTERRAINES DE STOCKAGE
M	ISNARD	VERITAS
M	PIGNALET	SOLETANCHE
M	PLUMELLE	CEBTP
M	SCHREIBER	SOLETANCHE

Appendix E

238

2,17E+09	2,17E+09		
2,05E+09	2,05E+09		
2,07E+09	2,07E+09		
1,98E+09	1,98E+09		
2,11E+08	2,11E+09		
1,95E+09	2,23E+09		
2,23E+09		2,10E+09	
2,31E+09		4,24%	

3127

6,12E+08	6,12E+08		
8,00E+08	8,00E+08		
5,56E+08	5,56E+08		
2,14E+09	2,14E+09		
9,55E+09	6,55E+08		
1,54E+08	7,96E+08		
6,55E+08		6,06E+08	
7,96E+08		65%	

261

2,18E+09	2,18E+09		
2,00E+09	2,00E+09		
2,07E+09	2,07E+09		
1,93E+09	2,00E+09		
2,00E+09	2,00E+09		
2,25E+09	2,21E+09		
2,00E+09		2,08E+09	
2,21E+09		4,62%	

3132

2,34E+09	2,34E+09		
2,57E+09	2,34E+09		
2,34E+09	2,54E+09		
2,54E+09	2,38E+09		
2,38E+09	2,48E+09		
2,26E+09	2,46E+09		
2,48E+09		2,42E+09	
2,46E+09		3,40%	

© The Editor(s) (if applicable) and The Author(s) 2024
J.-J. H. Rincent, *Ground Anchors*, https://doi.org/10.1007/978-981-97-4414-5

138				5125		
	2,12E+09	2,12E+09			2,98E+09	2,98E+09
	2,29E+09	2,29E+09			3,22E+09	3,22E+09
	2,02E+09	2,19E+09			2,51E+09	2,96E+09
	2,19E+09	2,24E+09			2,96E+09	2,91E+09
	2,30E+09	2,26E+09			2,91E+09	3,15E+09
	2,24E+09	2,17E+09			3,15E+09	3,25E+09
	2,26E+09		2,21E+09		3,34E+09		3,08E+09
	2,17E+09		2,86%		3,25E+09		4,74%

450 and 451 main text				5138		
					3,69E+09	5,19E+09
					5,19E+09	5,98E+09
					5,98E+09	5,21E+09
					5,21E+09	5,62E+09
					5,62E+09	4,91E+09
					7,57E+09	6,71E+09
					4,91E+09		4,78E+09
					6,71E+09		11,77%

Appendix F

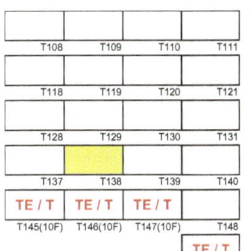

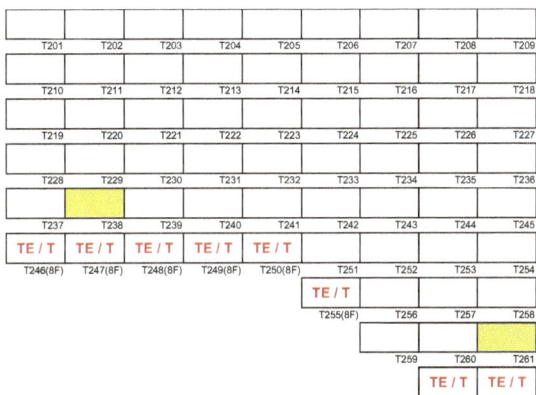

J.-J. H. Rincent, *Ground Anchors*, https://doi.org/10.1007/978-981-97-4414-5

T306	T307	T308	T309	T310	T311	T312	T313	T314	T315	T316	T317
T323	T324	T325	T326	T327	T328	T329	T330	T331	T332	T333	T334
T340	T341	T342	T343	T344	T345	T346	T347	T348	T349	T350	T351
T357	T358	T359	T360	T361	T362	T363	T364	T365	T366	T367	T368
T374	T375	T376	T377	T378	T379	T380	T381	T382	T383	T384	T385
T391	T392	T393	T394	T395	T396	T397	T398	T399	T3100	T3101	T3102
T3108	T3109	T3110	T3111	T3112	T3113	T3114	T3115	T3116	T3117	T3118	T3119
T3125	T3126	T3127	T3128	T3129	T3130	T3131	T3132	T3133	T3134	T3135	T3136

| TE / T | TE / T | TE / T | TE / T | TE / T | TE / T | TE / T | TE / T |
| T3142(8F) | T3143(8F) | T3144(8F) | T3145(8F) | T3146(8F) | T3147(8F) | T3148(8F) | T3149(8F) |

| T3150 | T3151 | T3152 | T3153 |

| TE / T |
| T3154(8F) | T3155 | T3156 |

| TE / T |

T411	T412	T413
T424	T425	T426
T437	T438	T439
T450	T451	T452
T463	T464	T465
T476	T477	T478
T489	T490	T491
T4102	T4103	T4104
T4115	T4116	T4117
T4128	T4129	T4130
T4141	T4142	T4143

118(8F)

T501	T502	T503	T504	T505	T506	T507	T508
T514	T515	T516	T517	T518	T519	T520	T521
T527	T528	T529	T530	T531	T532	T533	T534
T540	T541	T542	T543	T544	T545	T546	T547
T553	T554	T555	T556	T557	T558	T559	T560
T566	T567	T568	T569	T570	T571	T572	T573
T579	T580	T581	T582	T583	T584	T585	T586
T592	T593	T594	T595	T596	T597	T598	T599
T5105	T5106	T5107	T5108	T5109	T5110	T5111	T5112
T5118	T5119	T5120	T5121	T5122	T5123	T5124	T5125
T5131	T5132	T5133	T5134	T5135	T5136	T5137	T5138

Appendix G

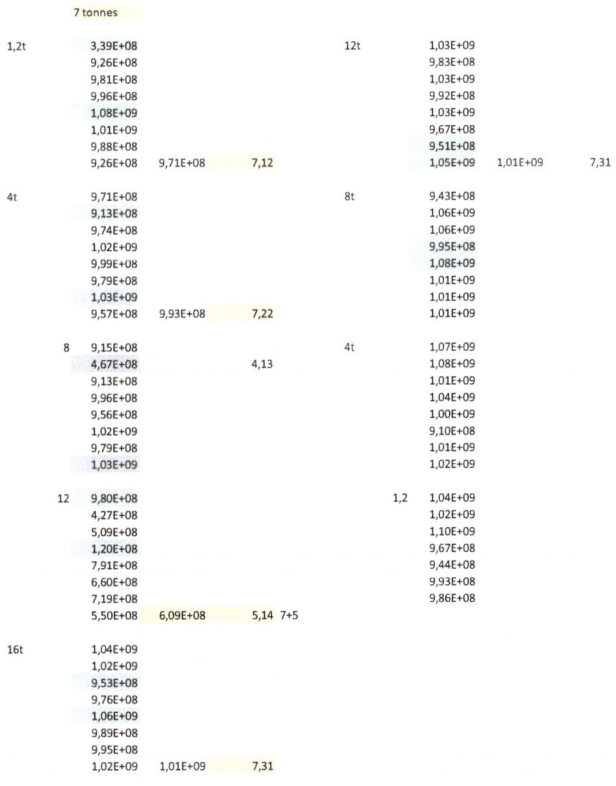

J.-J. H. Rincent, *Ground Anchors*, https://doi.org/10.1007/978-981-97-4414-5

Appendix H

Tie rod 412

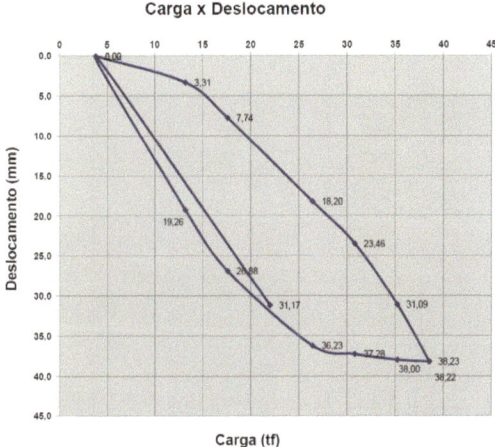

Source: Rincent BTP – Recife

Tie rod 504

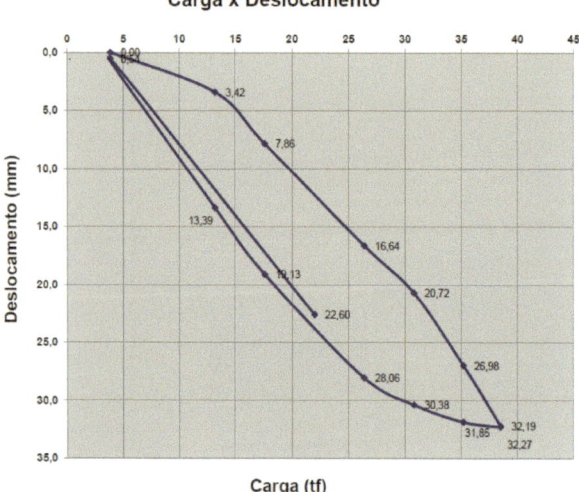

Source: Rincent BTP – Recife

Tie rod 567

Carga x Deslocamento

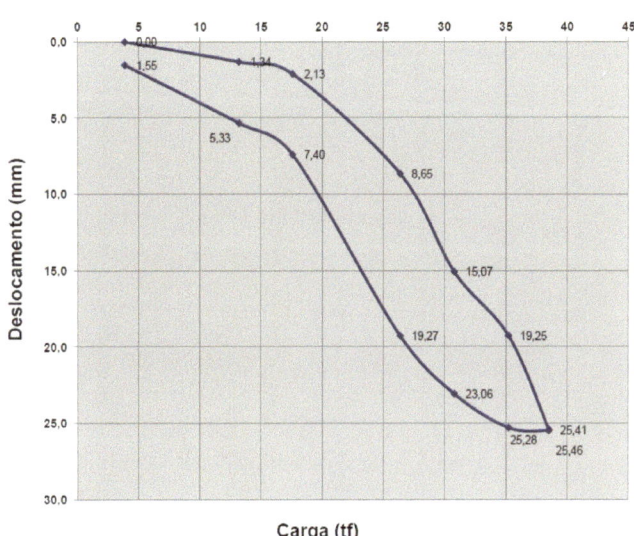

Source: Rincent BTP – Recife

Tie rod 580

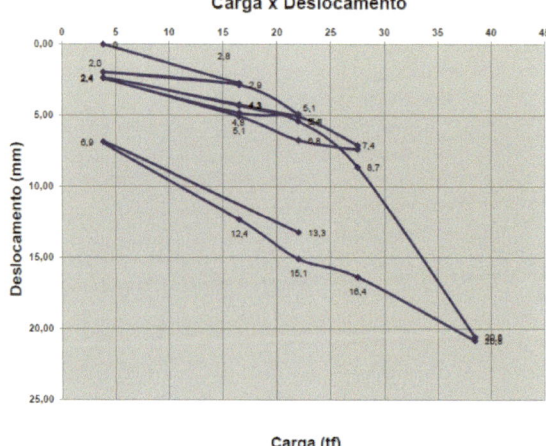

Source: Rincent BTP – Recife

Bibliography

Paquet J (Mai 1968) Etude vibratoire des pieux en béton, réponse harmonique- Annales I.T.B.T.P.

Davis AG, Dunn CS (1974) From theory to field experience with non-destructive vibration testing of piles. Porc.inst.civ.ingrs

Davis AG, Robertson SA (1975) Vibration testing of piles. Struct Engrg, London, June, A7

Paquet J, Briard M (mars 1976) Contrôle non destructif des pieux en béton—Annales de l'ITBTP n° 337

Guillermain P (Mars 1979) Contribution à l'interprétation géotechnique de l'essai d'impédance mécanique d'un pieu. PMC

Davis A, Guillermain P (aout 1979) Interprétation géotechnique des courbes de réponse de l'exitation harmonique d'un pieu. Revue française de géotechnique n°8

Norme Française (1994) NF P 94 160-4 Sols: reconnaissance et essais Auscultation d'un élément de fondations partie 4 méthode par impédance

Knothe K, Yu M, Illias H (2002) Measurement and modelling of resilient rubber rail-pads System. Dynamics and long-term behaviour of Railway Vehicles Track and Subgrade, Concluding Colloquium of the DFG-Priority-Program Stuttgart

Soyez L (2011) Contribution à l'étude du comportement des ouvrages de soutènement renforcés soumis à des charges d'exploitation ferroviaires. (Charges dynamiques et cycliques)

Rincent JJ (2006) Method and device for determining the tensile stress exerted on a sealed element. Patent OMPI n° WO 2006/010830 A1

Asli C (2008) Analyse des mesures et combinaison des raideurs dynamiques IUP EVRY

Mitaine L, Rincent JJ (2015) Innovative non-destructive technique for determining tension in ground anchors Revue Paralia Volume 8 pp s01.1-s01.7

Wilquin F, Horb C, Feng ZQ, Porcher G (2016) Dynamic non-destructive evaluation of rock anchorage. 3rd international Rock Slope Stability Lyon

Horb C, Saurel J, Rincent JJ (2021) Preventive maintenance and non-destructive monitoring of rock anchorages Tools resulting from dynamic tests and their correlation with static tests. 5E Stabilité de la pente rocheuse RSS Symposium, Chambéry, Norme suisse SIA 267/1 2013 Géotechnique ASTRA 120005 Tirants d'ancrage Edition 2022 V1. 13

Brevet P, Olivié F, Guilbaud JP, Raharinaivo A (1970-2000) Microstructure et propriétés mécaniques des a,ciers pour câbles Synthèse des travaux du LCPC Plasticité et endommagement

Piron A, Chulliat O, Morel F (2012) Suivi de la tension résiduelle par auscultation des tirants d'ancrage du parc hydraulique EDF

CFBR Comité français des barrages et réservoirs (29 janvier 2015) Symposium du CFBR à Grenoble Contrôle de la tension des tirants

ASN Autorité de sûreté nucléaire CODEP—DCN 2015—0 17985 Tirants d'ancrage précontraints

Poineau D (2010) Méthodes courantes d'évaluation structurale—Évolution des armatures et procédés de précontrainte. Extrait d'un document interne du SETRA

© The Editor(s) (if applicable) and The Author(s) 2024   151

J.-J. H. Rincent, *Ground Anchors*, https://doi.org/10.1007/978-981-97-4414-5

Ancrages TA (2020) RÈGLES PROFESSIONNELLES concernant la conception, le calcul, l'exécution, le contrôle et la surveillance

Porto Bomjardim T (2015) Ancoragens em solos comportamento geotécnico e metodologia

Ladiges S, Wark R, Rodd R (2016) Maintenance and Testing of Post-Tensioned Anchors for Dams and Appurtenant Structures

References

Ground anchors—Non-Destructive Testing NDT

The technique used to determine the tension force of ties has been in use since 2000 and has been applied to more than 10,000 tie rods. Considering that we perform 8 tests per tie rod, we have a database of 80,000 tests.

2002 Hertzian pylon stabilized with prestressed vertical ties.

2003 Lock at the port of Le Havre

Source: Rincent BTP France

2006—2007 Nuclear power plant cooling water feed channel, multi-year contract.

2008: Port of Brest (France) dock structure. Tension control of the tie rod to ensure the stability of the dock against waves. Multi-year contract.

© The Editor(s) (if applicable) and The Author(s) 2024

J.-J. H. Rincent, *Ground Anchors*, https://doi.org/10.1007/978-981-97-4414-5

Source: Rincent BTP France

Wuppertal, Germany,   Natal, Brazil

Source: Rincent BTP France - Rincent BTP Recife

2009: Bilbao, Spain—tunnel exit.

2009—2010: Retaining wall—Measurement of load increase as a function of excavated soil evacuation, new metro. Rennes

Source: Rincent BTP France

2011 Port of Calais   Retaining wall North of Paris

Source: Rincent BTP France

2012: Port of Nantes Tunnel Factory of Revin—Electricity de France

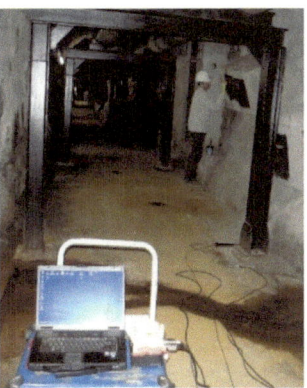

Source: Rincent BTP France

2014: Quay of the port of a Pacific island Ministry of the Armed Forces Reinforcement of the ground city of Meaux

Source: Rincent BTP France

2015: Barrage—Tie rod with a force of 700 tons (2020: second intervention)

Source: Rincent bTP France

2016 Abidjan, Ivory Coast—Property wall
2018: Tasmania, South Australia dam.

Source: Rincent BTP France

2019: Gramados, Brazil—Highway **750 tie rods**

Souce: Rincent BTP Recife

Fixing Mont Blanc cable car posts

Source: Rincent BTP France

2020: Railway, Brazil—**1,700 ties**. Alp's cliff reinforcement

Sources: Rincent BTP Recife - Rincent BTP France

2021: Barrage Minas Gerais Brazil Quais Caribes

Sources: Rincent BTP Recife – Rincent BTP France

2021 Barrage Etat of São Paulo

Source: Rincent BTP Recife